Your Brain and Your Self

Jacques Neirynck

Your Brain and Your Self

What You Need to Know

Translated by Laurence Garey

 Springer

Jacques Neirynck
Swiss Federal Institute of Technology
Lausanne
Switzerland

Translated and updated by
Laurence Garey

This book was translated from the French original:
Tout savoir sur le cerveau et les dernières découvertes sur le Moi
By Jacques Neirynck
Editions Favre SA, Lausanne
2006

ISBN 978-3-540-87522-2   e-ISBN 978-3-540-87523-9
ISBN (My Copy): 978-3-540-87538-3

DOI: 10.1007/978-3-540-87523-9

Library of Congress Control Number: 2008935327

© 2009 Springer-Verlag Berlin Heidelberg

Cover design: WMX Design GmbH, Heidelberg, Germany

Printed on acid-free paper

9 8 7 6 5 4 3 2 1

springer.com

For Christiane

If the brain was simple enough to be intelligible, we would not be intelligent enough to understand it.

Anonymous graffiti

Preface

Jacques Neirynck and I arrived in Lausanne at about the same time, respectively in 1972 and 1973, he at the Swiss Federal Institute of Technology in Lausanne (EPFL) and I at the Faculty of Medicine of Lausanne University. I was there to teach anatomy and carry out research on the brain. He pursued a very varied career: engineer, researcher and teacher, but also journalist (including a reputed consumer advice program on Swiss television), author (both of science and fiction), and politician.

I was very pleased when he agreed for me to translate the book he had just written: *"Tout savoir sur le cerveau et les dernières découvertes sur le Moi"*. The book had been inspired by a number of discussions with neurologists and biologists at the newly created Brain Mind Institute of the EPFL.

In this translation, the concepts and descriptions are preserved, although some of the original interview material has been abbreviated to try to make the text flow and to impart a more "international" feeling. This in no way minimizes the essential contributions of the interviewees

Throughout I have tended to use the masculine pronoun, but of course the feminine version is just as much implied. Equally, the generic "man" is used rather than try to find a more cumbersome alternative. We are here dealing with biology, not political correctness.

August 2008

Laurence Garey
Perroy

Contents

Introduction

How does my brain work? Why am I conscious? Where is my memory? Is what I perceive around me reality or just an illusion? We all ask these questions, which we could sum up in a single question: who am I? How is it that I have memories and that I feel I exist? What does it mean that my mind is free in time and space, and yet I am imprisoned in a body that is doomed to disappear? What happens to my mind when my body disappears? What are the risks of my suffering from a brain disease? Could my whole being eclipse because of a disease in which my body survives but my mind ceases to exist? What remedies are there? What hope does research hold out? Recent discoveries about the brain allow us to ask such questions more pointedly, hoping to define more clearly the relations of the brain with the mind, of man with his body.

I see this book as an "instruction manual" of myself. Our brain is an essential body organ, for the slightest defect in its function prevents us from using the others properly, even if they are healthy. Without a properly functioning brain, my "self" is deranged and may be completely destroyed.

If we wish to look after our digestive or cardiovascular systems we take care to exercise, pay attention to our diet and take medication against hypertension or cholesterol. That is all important, but we must also realize that our brain deserves just as much, if not more, attention. We need mental hygiene, just as much as bodily. Abuse of drugs or medication, intellectual or spiritual barrenness, mental idleness, cultural sterility, all leave indelible traces in our brain and, therefore, in our mind.

This book hopes to guide the reader, irrespective of their knowledge of science. The general public is often unaware of the quality and importance of today's research. Now that infectious diseases are largely mastered, although some menace us again, and cancer is more and more under control, diseases of the nervous system are becoming one of the principal causes of invalidity and death.

Nevertheless, a feeling of mistrust, even hostility, toward cutting edge medicine is marking certain political decisions. So, it is important to inform public opinion about the organ that differentiates human beings from other animals: about its fragility, about its accessibility to proper care, and about the research that is both possible and necessary. In this way people will be able to make informed decisions when called upon to comment on scientific matters as part of the democratic process.

J. Neirynck, *Your Brain and Your Self: What You Need to Know*,
© Springer-Verlag Berlin Heidelberg 2009

This book is based on numerous discussions with specialists, and not on the opinion of the author. It attempts to determine the state of the art. There are few notes or references to intimidate the reader to simply shut it. It is organized in chapters that can be read in continuity, but it is equally possible to discover the chapters in a different order or to skip any which may be of less interest. The reader's absolute right is to not read everything! For instance, as an appetizer he could read the case histories that punctuate the chapters. After a description of the brain and the tools used to investigate it, we progress to a consideration of four major pathologies, the diseases of Parkinson and Alzheimer, then stroke and tumors. We next reflect on those strange states of consciousness that are the "out of body" and "near death" experiences. We end by exploring current openings in the research on simulating brain function.

This venture into the world of neuroscience began by concentrating on the very special centers of excellence provided by the university milieu of Lausanne and Geneva. At the EPFL (the Swiss Federal Institute of Technology in Lausanne) a Brain Mind Institute was created in 2002, based on a novel fusion of technology and medicine, of engineers and biologists. The former are learning to study the phenomena of life, and the latter to exploit the artificial brain to penetrate the mysteries of the natural brain.

I want to thank all those who gave me their time, especially Patrick Aebischer, Olaf Blanke, Julien Bogousslavsky, Stephanie Clarke, Nicolas de Tribolet, Touradj Ebrahimi, Wulfram Gerstner, Rolf Gruetter, Pierre Magistretti, Henry Markram and Jean-Philippe Thiran, as well as to Sylvie Déthiollaz, Andrea Pfeifer, Béatrice Roth and Bernard Roy, for providing me with very useful documentation.

Chapter 1
The Controversial Seat of Myself

"As human beings we are possessed of a fundamental desire to understand, on the one hand who we are, and on the other where we come from. Two disciplines, neurology and cosmology, have strived to clarify these two metaphysical questions. Until recently only philosophers tackled the first of these, that of consciousness, and they approached it exclusively by using the capacity of their own minds. But how can we understand objectively that most subjective part of ourselves? For a long time philosophical musings have given us contradictory answers, between which it has been impossible to choose on a rational basis. What has changed today concerns the tools we possess for measuring and investigation, which allow us to begin to formulate these questions more rigorously. As specialists in natural science, we cannot claim to provide an unequivocal reply to a metaphysical question, but we can offer to ask the question better, in order to avoid wrong answers."

This scientific profession of faith was made to me by Patrick Aebischer, President of the EPFL one morning in November 2005 as I questioned him about the creation of the Brain Mind Institute, the existence of which in the heart of an engineering school was a surprising novelty. The aim of technology is usually to make life easier, not to penetrate its secrets. Now technicians are beginning to be interested in that most hermetically sealed part of nature, the human being. What can they offer that is new? What is the concrete benefit of their research? Can we better understand how we function by penetrating the very depths of our personality? Will engineers build more highly performing computers by imitating the biological systems of our brain?

Within the confines of our skull resides our whole being, a succession of thoughts and feelings to which each individual has sole access, if we exclude phenomena such as telepathy or clairvoyance that some claim exist. We perceive the outside world from our inside world, that is distinct from it and which stirs up a host of memories, prejudices, habits, emotions and desires. This intimate universe is not subject to the constraints of time and space of the outside world. We can project ourselves into the past or the future, revisit all the places that we have known and imagine those we wish to discover. We can invent a fictive world of stories that never happened but which seem larger than life to us. We can create sculptures, pictures and songs that charm us and which stem from the depths of this bottomless pit that we call ourselves. We have a sensation of freedom from the constraints of our body.

J. Neirynck, *Your Brain and Your Self: What You Need to Know*,
© Springer-Verlag Berlin Heidelberg 2009

This consciousness makes us human, but where does it come from? Consciousness is like time. We experience both so intimately and so forcefully that we fail when we try to define them. Swept up by time we can only watch it pass but never stop it to examine it at leisure. Equally, as we perceive our existence through our consciousness we cannot observe it with detachment.

The Dualist Fantasy

The common phantasm that we maintain on this subject divides our being into two distinct parts, body and soul, brain and consciousness. It is that favorite temptation of western culture, dualism: material without a soul, and an immaterial soul. The most radical formulation of this doctrine was by French philosopher, Henri Bergson (1859–1941) in his work on *Matter and Memory* (1896) that "the hypothesis of equivalence between psychological state and cerebral state implies a downright absurdity". In other words, the feelings we experience, the emotions we undergo, the intuitions from which we benefit, the memories we recall, all, according to him, are independent of any material substrate in our body. On the contrary, they are separate, manifesting themselves in some immaterial entity, invisible and spiritual, which people commonly call the soul. This is a naïve translation of the impression that we all feel of inhabiting a body which is distinct from our consciousness.

Once we have made this mistake of separating our being into two radically different parts, that are contradictory, opposite and irreconcilable, we shall have a hard time trying to put ourselves back together again. When he needed to reconcile the two, the much earlier French philosopher René Descartes (1596–1650) selected the pineal, a tiny organ buried in the middle of the brain, as the seat of the soul within the body. As a seventeenth century philosopher, he had just one reason to imagine this colocalization: the brain is symmetrical (well, almost!), and divided into right and left hemispheres, apart from certain structures such as the pineal. Today we know about its real function in synchronizing sleep with the alternation of light and dark. Certainly an important function, but rather modest compared with what the philosopher had decreed with the unbelievable impudence of an intellectual who is the reprehensible victim of a trick played on him by his own brain, that he imagines he can understand purely by reflection.

For Bergson, Descartes and, much earlier, the ancient Greek philosopher Plato (424–348 BC), all of whom were dualists, the soul is not localized within the brain, but is rather a separate entity, immortal by its very nature and thus immaterial. When a man dies, his soul leaves his body for another world. As a bottle is a container for wine, the body is a sort of fortuitous, ephemeral shell which contains the imperishable part of our being. Western values are marked by an inexorable confusion about the existence of the soul and its immortality. To distinguish between brain and soul reflects a metaphysical option which interferes with scientific investigation. When a neurologist reveals that a psychological condition corresponds to a state of the brain, he seems to undermine the very foundations of our civilization.

To avoid any misunderstanding, it is useful to mention that dualism represents a philosophical attitude, but it does not represent a religious belief, in spite of the similarities of the concepts, which often leads to some confusion. In the Christian faith, body and soul are considered as inseparable, so much so that the Creed does not profess immortality of the soul, but the resurrection of the body. Nothing is more instructive in this context than to consult the articles on the soul in the Catechism of the Catholic Church, published in 1992. In this work, written by theologians for theologians rather than for the grass-root faithful, we find the definition: *Body and soul but truly one. Man, though made of body and soul, is a unity. Through his very bodily condition he sums up in himself the elements of the material world. Through him they are thus brought to their highest perfection.*

In the same way, Paul Ricoeur, the French humanist philosopher, in discussions published in 1998 with Jean-Pierre Changeux, the well-known molecular neuro-biologist at the Collège de France and Pasteur Institute in Paris, clearly made the distinction between dualism and spiritualism. The spirit does not cease to exist if we consider that it is linked to the body.

In antiquity people went so far as to weigh a dying person to determine if the departing soul could be detected by a change in weight. This absurdity reveals the true nature of dualism: to attempt to demonstrate materially the existence of the immaterial. As if the latter could only exist by derogation, on the fringes of the observable, as if what cannot be measured did not really exist, as if the invisible only existed by virtue of its visible manifestations. Even today people publish books and articles attempting to localize the soul in a part of the brain, or in some cerebral function.

This dualist phantasm is not without practical consequences. We shall consider three, in criminal law, in psychiatry and in medicine.

Insanity and Criminal Responsibility

In the application of criminal law it is usual, with the help of psychiatric expertise, to distinguish between an irresistible urge and a voluntary act. Certain criminal behaviors might be attributed to the mechanical action of a sick brain, whereas others are the result of a deliberate, considered decision under the authority of a sovereign will.

The first are excusable to some extent, as the criminal is considered a sort of robot, without free will. He should be locked up simply to prevent him doing more harm, and certainly not to punish him, for he is not responsible, because he cannot stop himself committing reprehensible acts. To force the comparison, one could say that he is a body without a soul.

Crimes of the second sort are punished in the name of the moral order which controls the actions of men with a free will. This supposes that the accused can make decisions independently of the state of his brain, and that he can control his brain by means of some immaterial mechanism. If he commits a crime he must be imprisoned punitively. He has a soul that is free of bodily impulses, and it has failed in its ethical responsibilities.

For inexplicable reasons some men seem to be devoid of this moral mechanism, which functions perfectly in others. Experience shows that experts who have to make such evaluations are mistaken with a disturbing frequency. The insane are imprisoned and finish by committing suicide, while hardened criminals are released and inevitably commit new crimes. An understanding of the brain may help to escape from this impasse by demonstrating that we are all less free that we may imagine and that, in the end, free will is perhaps pure fiction.

Nervous Diseases, or Psychological Diseases?

The same is true in psychiatric practice. We continue to distinguish between neurological and psychological diseases. On the one hand we recognize brain damage as a result of an accident, or infection by the prions of Creutzfeldt-Jakob disease, or the viruses of meningitis, or the bacteria of encephalitis; we accept that multiple sclerosis and the diseases of Alzheimer and Parkinson are due to dysfunction in specific parts of the brain. On the other hand we have autism, chronic depression, manic depression, anxiety, schizophrenia and diverse phobias that are diseases of the mind, curable by a simple series of psychiatric consultations, assuming that the patient really wants to be cured. Indeed, his symptoms are due to some spiritual weakness that he can, and therefore should, overcome.

The former diseases are the result of biological fatality that the patient has to endure without being in any way to blame: he is a pure and guiltless victim. In contrast, the latter troubles are the direct responsibility of the patient, especially when there is a question of interpersonal relations. The patient has become intolerable as a result of his own free will. He thus endures the double burden of suffering from his condition, and from the pain of his own feeling of responsibility.

However, we can no longer accept such a clear distinction between the neurological and the psychoanalytical. Their interrelations must be seen in the new light of recent discoveries, especially those concerning medical imaging of what goes on inside our heads.

Western Medicine and Cartesianism

For three centuries the aim of western medical research was to understand the physiology and pathology of the human body, including the brain, an organ like any other. Research on the mind was transposed to the realms of theology, philosophy and, later, psychology. The human body was seen as an object devoid of sentiments. The relationship between a disease and the psychological state of the patient was only taken into consideration at a late stage, and then only partially. The dialog between doctor and patient was reduced to a sort of police enquiry, as objective as possible, that led to an automatic prescription of medication, and the performance of biological analyses and technical acts such as surgery.

We are now just beginning to admit that psychological problems can influence bodily disease, and *vice versa*. Popular wisdom already taught that jealousy, envy, sadness, avarice, anger and solitude were sources of cardiac, digestive and dermatological disease, although in western medicine they had little interrelationship. Disease was treated chemically and surgically without our being aware that we were tackling the effects and not the cause, and that such dualist and materialist medicine has limited efficacy. This flawed conception of medicine, buoyed up by the growing specialization of medical practitioners and by an explosion of technology, diminished the quality of medicine and largely wasted economic resources.

Western medicine has created a chasm between body and mind by its insistence on the dualist fantasy. This has triggered a fascination for "alternative" medicines, often of oriental origin, including acupuncture, homeopathy and the laying on of hands, in all of which the practitioner takes the time to interact with the patient. This does not mean that these medicines are more effective overall than western medicine, but it does point to a fundamental deficiency of the latter. By supposing too often that human psychology has nothing to do with human physiology, we commit a profound error, that of neglecting the interaction of the brain with the rest of the body.

Before speculating about our mind, we should try to understand better that bodily organ, the brain, that we suppose to be its seat. Our body forms a unit: we do not intellectualize when we trap our finger in a door because the brain is set in motion by an overwhelming input. But if we lose a finger we do not lose our intellect. We must first describe the brain as revealed by the anatomist's scalpel and by the cerebral imagery of the scanner. Only later may we be able to grasp something of the indispensable, but confused, concepts that we call consciousness and memory, mind, spirit and soul.

Modern research on this physical organ obviously has little relevance to Bergson's dualist approach, but all that we have learned over the last century argues against his speculations. Today a brain specialist may legitimately rewrite what Bergson said, so that it becomes: "the hypothesis of equivalence between psychological state and cerebral state **has turned out to be remarkably fertile in neurological and psychological research**". This is a fact that can no longer be challenged. He could even defend a hypothesis that totally contradicts that of Bergson in that: "the hypothesis that there is **no** equivalence between psychological state and cerebral state implies a downright absurdity". But neurologists are allergic to this sort of aggressive declaration. They do not speculate, they observe. As specialists they know the brain's fragility. So they use their own with the greatest caution.

The History of Errors About the Brain

Real advances in our knowledge of the brain are a recent phenomenon: only in the last two centuries has brain research advanced in an explosive fashion. For many centuries before that our gray matter remained hidden within our skull, and we dared not touch it for fear of the retribution of the ecclesiastical, academic and political authorities of the moment. It took a certain ideological progress for researchers to conceive that the brain might not work as a single whole, but that it fulfilled numerous functions such as audition, vision, speech, touch, pain and movement, all localized in specific regions.

Many relics of the past, such as bones, tools and drawings, suggest that our ancestors had suspected that the brain was a vital organ. Among other things, archaeologists have found evidence of surgical operations, such as trepanations, in very diverse cultures. But it is not enough to examine a brain removed from a dead skull, or even to dissect it, to understand its function. To know more we had to exceed the limits of our own perception. We had to await the dawning of a new technological era that saw the development of instruments that performed better than the human eye, and even allowed us to see the inside of the skull in life.

Our ancestors had serious disagreements about the primary function of the brain. "Men ought to know that from the human brain and from the brain only arise our pleasures, joys, laughter, and jests as well as our sorrows, pains, grieves and tears," said Greek physician Hippocrates (460–379 BC) at the dawn of western science. This founding father of science had rightly guessed that the brain is the center of sensation and the site of intelligence.

On the contrary, the Greek philosopher Aristotle (384–322 BC) was persuaded that the heart was the centre of the intellect, by means of some highly significant philosophical regression. For him, the brain was nothing more than a piece of thermal mechanics to cool the blood as it was overheated by emotions felt by the heart. The Aristotelian concept of the brain was of a radiator. In our present-day colloquial language we still speak of the heart as the center for the emotions, as opposed to the brain, the center of the intellect. Let us not forget what Shakespeare said in *The Merchant of Venice*: "Tell me where is fancy bred, or in the heart, or in the head?"

The Greek physician Herophilus (335–280 BC), was considered the father of Anatomy. He noted that every part of the body was connected to the spinal cord by discrete nerves, the course of which he was able to trace. Another slightly later Greek physician, Galen (131–201), described the brain as composed of two distinct parts, the encephalon responsible for sensations, and the cerebellum that seemed to control the muscles. Galen also supported a theory of vital humors, fluids that circulated throughout the body, including the brain. The nerves were hollow tubes in which circulated the same humors, transformed into psychic pneuma in the center of the brain itself in the cavities that we call the ventricles. Sensation and movement resulted from a judicious mixture of these four humors. In the seventeenth century this theory was beginning to look ridiculous, as we see in some contemporary stage comedies, notably by Molière in France.

At that time, no-one could conceive that nervous signals might circulate and be processed in the form of electrical impulses, for the physical phenomenon of electricity was not studied until the eighteenth century. Nevertheless, the theory of humors was not entirely false: electrical impulses traveling between neurons cross gaps at the synapses where they are relayed by chemical transmission thanks to molecules called neurotransmitters. And indeed we now know that nerves are in a way "hollow", with a flux of substances in both directions, the anterograde and retrograde axonal flow. Further, the brain also secretes hormones that profoundly influence our behavior.

In particular, the endorphins attenuate the sensation of pain and instill a sensation of well-being. So, we must accept that information is transmitted and handled by both electrical and chemical means.

The advancement of ideas about the brain was blocked for almost a millennium by the controversy between Hippocrates and Aristotle. At the time it was forbidden to touch the human body and one had to await the Renaissance for certain pioneers to dare to dissect a human cadaver. The Italian master of art and science, Leonardo da Vinci (1452–1519), and Brussels-born Andreas Vesalius (1514–1564), two major pioneers of human anatomy, published numerous detailed drawings of the human body (Figure 1). People began to grasp that cerebral function was localized in the brain substance itself, rather than in the liquid that bathes it, although both these masters still illustrated, and put a certain emphasis on, the role of the ventricles.

Further, the great French philosopher René Descartes (1596–1650) continued to support a theory of animal spirits, although adding an extra dimension that would for long nourish the controversy. According to him, it was impossible for the human soul to result from a mechanical process. He perfected his theory of dualism, of body and soul, proclaiming a separation of physical and mental functions. The former depended on the flow of humors, the latter on God, with an interface between the two, supposedly at the level of the pineal, as we saw above.

Happily, toward the end of the seventeenth and beginning of the eighteenth centuries, certain scientists, less obsessed by metaphysics than by observation, turned their attention in more detail to the structure of the brain. They noted that there seemed to be two types of brain substance: a gray matter (like the beloved "little grey cells" of Hercule Poirot, the hero of so many of Agatha Christie's novels),

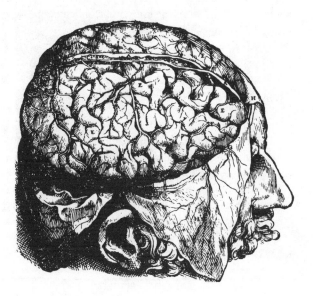

Figure 1 Drawing by Vesalius (1543) of the cortical gyri

draped over a white matter. Further, they also noted that the major fissures and gyrations of the outermost gray matter, the cerebral cortex, were somewhat similar in most brains. This left but one step to take to a description of cortical localization, the division of the cortex into functionally specialized areas.

In the middle of the eighteenth century, physics underwent its golden age with the theories of English physicist Isaac Newton (1643–1727) followed rapidly by the development of electromagnetism. In Italy Luigi Galvani (1737–1798) showed that it was possible to cause a contraction of the muscles of a frog's leg by the application of an electrical discharge (Figure 2). Even if it seems trivial today, this elementary experiment opened new perspectives. Little by little the vision of nervous transmission by means of humors was eclipsed by a theory based on electrical transmission.

In France François Magendie (1783–1855) demonstrated at the beginning of the nineteenth century that, at the level of the spinal cord, nerves divide into two roots, one responsible for motor information to the muscles and the other conveying sensory information such as touch and pain, hot and cold. This was a step toward a concept of specialization of the different structures of the nervous system. To test such a theory, researchers ablated different parts of the brain of animals and observed the effect of these localized lesions.

The German physician Franz Joseph Gall (1758–1828) wondered if the various folds visible on the surface of the brain could also be implicated in separate functions. However, instead of limiting himself to the brain itself, when he would have been right, he invented a new science, that he called phrenology, based on the idea that one

Figure 2 Galvani's laboratory where he proved the electrical nature of nerve conduction

could determine a person's traits of character by studying the bumps on the **outside** of the skull (Figure 3). This novel theory is erroneous, but has the merit of being the first to hint at the existence of localization of function in the cortex. Phrenology subsisted into the twentieth century, and in 1907 people were conscientiously measuring the bumps of the skull with an electric phrenometer!

Gall's theory was erroneous on two counts. First, although the areas of the cortex are indeed specialized, these areas do not correspond to traits of character, such as sexual instinct or combativity, poetic talent or metaphysical sense. Today we know that one's mental capacity depends on interaction between different areas each with much more operational functions. Further, the size of a functional area is not

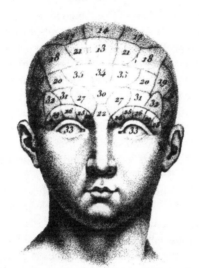

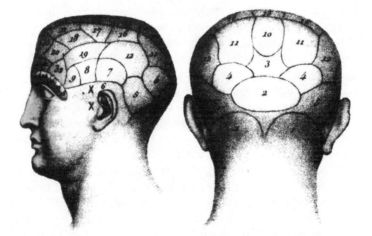

Figure 3 Gall's phrenological maps in which he designated 27 mental faculties

proportional to the degree of development of this function, as Gall had naïvely believed. So, the study of bumps cannot reveal anything about the mind. But Gall had the merit to affirm, like Hippocrates, that the brain is the organ of the mind, thus renouncing dualism unequivocally. Such an option carried a heavy responsibility: an individual was intelligent or stupid, affectionate or frigid, innocent or guilty, depending on the structure of his brain.

Finally, it was the observations of the French neurologist Paul Broca (1824–1880) that convinced scientists that the various functions of the brain were localized in precise regions. Broca described the case of a patient who understood language but who could not speak. After his patient's death Broca discovered a lesion in a specific part of his cerebral cortex and concluded that language production must be intimately related to that area. We still refer to this speech area as Broca's area. Later, many animal experiments reinforced this developing concept of cortical localization.

At around the same time, one of England's greatest naturalists, Charles Darwin (1809–1882), elaborated his theory of the evolution of species and natural selection by the survival of the fittest. His theory opened new perspectives for experimental brain research: it became possible to rely, at least to some extent, on animal models that could be extrapolated to man. The concept underlying this approach was that the structure and function of the brains of all animals, including man, were derived from a common ancestor. Species shared common characteristics, while differences resulted from the adaptation of each species to its natural environment.

This stage of scientific thought had profound consequences because it affirmed the unity of living beings. Because man was descended (or ascended!) from other animals, his brain is the result of slow evolution, that is in no way magic, although the result is prodigious. As we shall see below, we can learn a lot about man's brain by studying the mouse's, for the basic component is the same: the brain cell or neuron. The differences are both quantitative and qualitative: man has more and they are organized differently. The same similarities and differences exist between a simple calculator and a powerful computer.

In 1839, while working in Belgium, the German physiologist Theodor Schwann (1810–1882) proposed a modern version of an older theory according to which living matter consists of microscopic units, the cells. The study of the way in which cells are "weaved" into tissues is called histology, and it developed by a happy coincidence of progress in optics (the development of good microscopes), and the elaboration of techniques allowing tissues to be fixed (preserved by precipitating their proteins), cut into very thin sections and stained, so that they could be visualized by microscopy. That there were similar cells in the brain had been postulated by scientists at least as far back as Marcello Malpighi (1628–1694) in the seventeenth century. But it was not until another Italian, Camillo Golgi (1843–1926) perfected a method to stain individual nerve cells that the real structure and diversity of the neuron was revealed. The Golgi method is based on silver impregnation of a small proportion of the total neurons in a piece of brain, but each is impregnated essentially completely, with all its protoplasmic processes or arborizations. However, Golgi make a mistake about the function of neurons when he concluded that they formed a continuous network, with the cell bodies like nodes in a complex electrical circuit, within which signals circulated in some mysterious fashion.

The masterful work of the renowned Spanish histologist Ramón y Cajal (1852–1934), using the Golgi method, definitively demystified the dogma of humors in their hollow canals, analogous to the blood circulation (Figure 4). He also convinced the scientific world that neurons did not form a continuous network, but

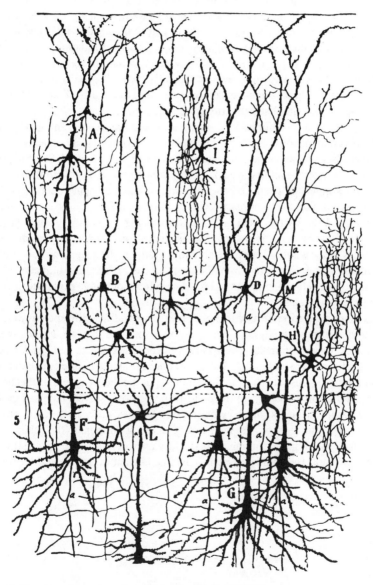

Figure 4 Drawing of a section of cerebral cortex by Cajal (1909)

were separate from each other, making contacts at specific minute regions. We now accept that these contacts are at the synapses, where chemical molecules cross the gap between the electrically excitable neurons. Neurons were finally acknowledged as the basic unit of the nervous system. Information enters the body of the neuron, is processed, and is then transmitted to other neurons that continue the good work. We can imagine, by much simplification, that they fulfill the same function as a logical circuit in a computer.

In 1906 Golgi and Cajal shared the Nobel Prize for Medicine, in spite of their conflicting theories, for at that time the idea of producing a wiring diagram of the connections of the brain seemed in itself so extraordinary that little attention was paid to what seemed a secondary controversy of how information was processed. Overall, Cajal's thesis is the right one, but with a few modifications, as we shall see later. But Golgi was also right in part. Today's reconciliation in the controversy between these two great scientists demonstrates that the reality of the brain is much more complex that simplistic theories might seem to suggest.

The brain processes information coming to it from our senses, and it controls our actions. If certain actions resulting from incoming information are repeated, the synaptic connections between the neurons concerned will be reinforced. So, synapses are the material support of all that we learn, from how to ride a bicycle to memorizing a text. In this simple (simplistic!) model the brain is like a computer that is capable, if not of programming itself, at least rewiring itself as a function of the tasks that it performs regularly. By repeating tasks, we learn to carry them out without having to think: part of our existence is governed by an automaton or robot, programmed by the experiences of the whole of our preceding life. Man is at the same time both sculpture and sculptor. Just as life sprang from the spontaneous organization of matter, thought springs from the self-organization of life. Herein lies a plausible explanation, but it is astonishing nevertheless.

This double question of why matter comes alive and why life spawns thought is by its very nature metaphysical. We can either imagine that we are dealing with pure chance, or we can make nature deterministic. This debate goes back two millennia, the Epicureans defending the former concept and the Stoics the latter. In the USA certain religious integrists still attempt to put the theory of evolution and creationism on the same footing: so the controversy is not resolved, and will probably never be for it touches on the very definition of our destiny. There is no objective answer to this question so we shall not consider it further. If the reader is still inspired by a vain hope of finding a satisfactory answer to this sort of question, then he should give up! He would do better to indulge himself in doubt and astonishment.

The story of Phineas Gage, the man with a hole in his head

Phineas Gage was a foreman for a railroad construction company, Rutland and Burlington, on a site in Vermont, USA. One summer day in 1848, Phineas was the victim of an accident that was spectacular both in its circumstances and its consequences.

He was working to blast a rock. To do this he had to drill a hole, fill it with powder (dynamite had not yet been invented), then cover the powder with sand, before tamping the mixture with an iron rod and lighting a fuse. Due to a mistake in his technique, Phineas tamped the powder rather than the sand, and it exploded. The tamping iron was propelled upward, penetrating Phineas' left cheek, the base of his skull and the front part of his brain before exiting the top of his head.

Phineas did not die from his injuries and was cared for in the local village by a certain Dr John Harlow. The patient remained conscious and perfectly lucid, and was able to both walk and explain to his doctor what had happened. He recovered in less than two months, but the consequences of the wound to his brain were astonishing. His senses were spared, apart from the total loss of vision in his left eye. He spoke without difficulty, and could walk and use his hands as if nothing had happened.

In contrast, his character had completely changed. He became aggressive and rude, obstinate and capricious, whereas before the accident he was considered to be perfectly well-balanced and well-behaved. His entourage considered that he was no longer the same man, even if he still possessed the body of Gage. His company refused to rehire him, for he had become useless in his former job, owing to his bad character and his incapacity to take rational decisions, in spite of his physical ability being intact.

Gage led a miserable life, emigrated to Latin America, worked as an agricultural laborer, and became a circus freak. He developed epileptic seizures and died in San Francisco in 1861. No autopsy was performed, but Dr Harlow obtained the consent of the family to exhume Phineas Gage's skull, which is now preserved, with the tamping iron, in the Anatomical Museum of Harvard Medical School.

More than a century later, Hanna and Antonio Damasio and their colleagues at the University of Iowa used modern medical imaging to make a virtual reconstruction of Phineas' skull and estimate in detail the damage that his brain had undergone. So we can determine that the lesion was essentially limited to the medial parts of the frontal lobes of both hemispheres adjacent to the longitudinal fissure between them. Phineas' selective deficits allow us to deduce that this region must be concerned with programming future behavior in accordance with learned social rules, and with making rational choices in life style. It includes important so-called association cortical areas which, if compromised, lead to the personality becoming fickle, asocial and aggressive.

Gage was Gage, but only as long as his frontal association cortex was intact.

Chapter 2
A Simple Architecture of the Brain

Let us take over where Golgi and Cajal left off. We can stain the cells of the brain, the neurons, and observe their ramifications.

The Neuron

The neuron is a cell, rather like the cells of the liver and the heart, but its function is completely different. It is not part of an organ that pumps or cleanses the blood, but its job is to receive, process and propagate information over a considerable distance. Before gripping an object we need to perceive it, localize it, and decide to grip it. So the result depends on visual perception and processing, followed by a motor command to our arm. All this is accomplished by neurons specialized in these various functions (Figure 5).

These specialized functions of a neuron have endowed it with a characteristic shape. It consists of a cell body, or soma, surrounded by several ramifications or processes, that can extend for a meter or more. Neurons form a complex network that allows the redistribution of information, in parallel or in series, to different parts of the CNS (the central nervous system, consisting of the brain and the spinal cord), and to the muscles via the peripheral nervous system. Information is relayed in two different ways, as we have seen earlier: electrical inside the neuron and chemical from one neuron to another (Figure 6).

Neurons possess two types of processes: usually several dendrites and a single axon. Put simply, dendrites receive information coming into the soma (afferent information), the soma summates it, and the axon transmits the resultant efferent signal to other neurons or to muscles. The axon only fires if the sum of the activity in the soma exceeds a certain threshold. A single axon ramifies to make typically around 10,000 contacts as information is processed further.

So, neurons use their highly specialized structure as much for receiving as for processing and transmitting signals. Each neuron receives information from thousands of other neurons and passes this information on to thousands of others. Information is relayed from one neuron to another by means of a chemical process, utilizing special molecules, the neurotransmitters, to cross a minute gap between

J. Neirynck, *Your Brain and Your Self: What You Need to Know*,
© Springer-Verlag Berlin Heidelberg 2009

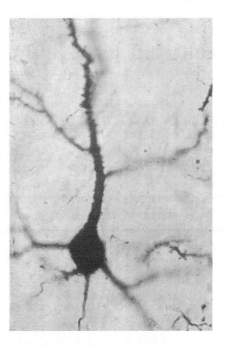

Figure 5   A neuron of the cerebral cortex stained with the Golgi technique. From the large cell body at the bottom stem several dendrites, with their spines, and a single axon from the extreme lower surface

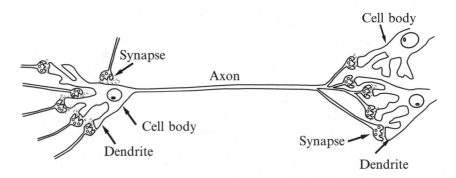

Figure 6   The electrical impulse is transmitted along the axon to the synapse

the terminal of the afferent neuron and the membrane of the next cell. We have already noted that this combination of afferent terminal, gap and the membrane of the next cell is called a synapse, a term coined by the renowned English neuro-physiologist Charles Sherrington (1857–1952) in 1897. As a chemical reaction is relatively slow compared to electrical transmission, the synapse delays conduction, which may at first sight appear paradoxical. Could this be an imperfection in the natural order?

Indeed, there exist axons and dendrites that are in direct electrical contact, at an electrical synapse that short-circuits the slow chemical relay of the chemical synapse. The aim of these electrical contacts is the rapid transmission of certain signals that are of capital importance for survival, or for certain reflexes that enable us to avoid sudden physical danger without having to think about it.

So, if it is possible to provide efficient electrical contacts, we are forced to wonder why chemical synapses, which seem at first sight less effective for signaling, exist at all. In fact, they play another, equally essential role: that of allowing learning and memory, for which the efficiency of the synapse must be able to evolve with time, rather than ensure instantaneous transmission.

Thus, this is not an imperfection of nature, but nature's ruse. In a computer, the memory is distinct from the processor: it is in the form of minute magnetic charges on a hard disk. In the brain there is no hard disk: memory consists of connections between elementary circuits that are, in essence, the neurons. It evolves continuously with time, and is widespread, not localized.

Very few neurons can actually divide in the mature brain, nor can they regenerate if damaged. This incapacity means that trauma, intoxication, or a deficit of oxygen due to cerebral vascular pathology interrupting normal blood flow will damage the CNS irreversibly. Brain diseases, that we shall consider later, are profoundly influenced by this characteristic of neurons: we dispose of a certain stock of neurons, mostly born during fetal life, and we slowly lose a significant number during adult life, estimated by some scientists as at least one neuron per second. This essential fragility is a feature of highly differentiated cells like those of the brain. Other organs, like the intestine, for example, are lined with cells that never cease to be renewed.

This does not mean that, in certain well determined circumstances, new neurons cannot be produced from precursors, the neural stem cells, put there for just that purpose. But this phenomenon remains the exception rather than the rule. To be sensible we must do all we can to preserve our own capital of neurons that we inherit at birth. At least, such is the classic wisdom of neuronal function. We shall add some nuance to this later for, after all, the concept is so simple that it may be the result of our desire to understand. Indeed, we can understand better if we simplify: by imagining a model that simulates nature we attain a certain degree of comprehension. But we must avoid ensconcing ourselves in this model. For example, for a long time we neglected the role of another type of cell found in the brain.

The Glial Cell

Glial cells are the other major component of the brain. For a long time they were considered to play a purely mechanical role "supporting" the neurons: glia is derived from the Greek word for glue.

Although not participating as obviously in neurotransmission as the neurons, glial cells play an important role in supporting neurons functionally, in terms of maintaining their ability to transmit signals. There are various types of glia in the brain, notably astrocytes, oligodendrocytes and microglia. The total number of glial cells is some ten times more than that of neurons. We might therefore have realized from the outset that they were not there by chance and that the evolution of the brain had not favored them without good reason, that is, without some essential function.

Glia are smaller than most neurons, and although they do have branch-like ramifications, these are neither axons nor dendrites. Among their clearly delineated roles one can cite in particular the modulation of the speed of propagation of nerve impulses and the supply of energy needed to maintain neuronal function. They also have a profound influence during development and maturation. Furthermore we suspect that glia might aid (or in certain cases might hinder) recovery after neuronal injury, and that they might play a role in several diseases, such as Alzheimer disease, multiple sclerosis, and schizophrenia among others.

Recently it has been determined that glia can also exchange information, not by the intermediary of connections between axons and dendrites, but by the production or uptake of neurotransmitters, or even by controlling the chemical composition of the fluid that bathes the neurons. Borrowing an image from telecommunications, we may consider the brain as, on the one hand, the equivalent of a telephone network with electrical connections (the neurons) and, on the other, a network of radio transmitters operating "wirelessly" over a distance (the glia).

But we do not have two transmission systems operating independently from each other. The glia supervise the function of the neural networks, not only at the synapse between two neurons, but over the whole extent of the conducting elements. An axon conducting electrical impulses releases into its environment molecules of ATP (adenosine triphosphate) that are detected by receptors on glia. This information influences certain genes that affect the way in which some glia (the oligodendrocytes) produce isolating layers of myelin along the axons, which ultimately controls the speed and efficiency of the transmission of the impulses. It is as if, in a computer made up of logical circuits, there existed a second control circuit whose job it was to improve the transmission of the digits 1 and 0 as they occurred.

The Fantasies of the Neuron

As a general rule electrical activity in a neuron travels in one direction. Electrical excitation of the dendrites, the input to the system, reaches the cell body where summation takes place and the resultant triggers activation of the axon if a defined threshold is attained. The axon is the output of the system. In the computer the basic component, the logical "AND" circuit, possesses two inputs and an output that has a value of 1 if the two inputs are 1. The temptation is great to imagine that the brain works like that, but in fact it is more complicated.

The neuronal electrical signal can propagate in both directions along a process, although the synapse is unidirectional. Furthermore, nerve processes are not simply elements of transmission in one direction or the other. In certain circumstances they can emit an impulse without having received one. They can also produce hormones that affect synaptic function, and can synchronize assemblies of neurons so they function as a system.

To ultimately complicate an already confused scenario, synapses are found not only at the junction between an axon and the dendrites of neurons further down the line. Electron microscopy has shown that synapses can also form on the neuronal soma and even between two axons or two dendrites. So neurons cannot be reduced to elementary processors, obediently performing summations and transmitting them further to other processors. It seems, rather, that they can be connected in a multidirectional fashion. In addition to what we have already discussed, assemblies of neurons can be coupled, not only by electrical impulse transmission or by neurotransmitter molecules, but also by electromagnetic interference, that is to say by radiotransmission, in the full sense of the term: all variable electrical currents create a magnetic field that, in turn, induces an electrical current in neighboring conductors.

At this stage we glimpse a possible explanation of coherent behavior of the human brain. Its activity is not comparable to that of the central processor of a computer treating information sequentially according to a rigid program. The brain reacts as a whole, even if certain regions are specialized. The specific capacity of man to integrate a mass of information into a coherent vision of his environment, the place he occupies within it and the activity that he can pursue there, all stem from cooperation between these regions.

Neurons and glia handle information at the microscopic level, but one cannot reduce the understanding of the brain to such an elementary level, just as a painting is not merely the sum of the pigments used, a symphony a sequence of notes and a poem a series of letters.

Indeed, the neurons are not connected haphazardly. Within the brain several quite distinct "organs" can be identified.

The Structure of the Nervous System

The CNS consists of the brain and the spinal cord, both bathed in a special cerebrospinal fluid (CSF). Weighing on average some 1400 grams, the brain has three main parts, the cerebral hemispheres, the cerebellum and the brainstem. There is no relationship between the very variable brain weight of individuals and their brain power. Lord Byron, at 2300 grams, and Emmanuel Kant (1600) are above average, Dante (1500) and Marilyn Monroe (1400) around average, and Anatole France and Franz Joseph Gall (both 1000) notably below average.

In spite of its great complexity the brain is, for the most part, composed of just the two cell types we discussed earlier, neurons and glia. This excludes such important supporting tissues as blood vessels. It contains more than 100 billion neurons (10^{11}) and around ten times more glia. That means there are about 70 million neurons per milliliter, that is the size of a dice measuring a centimeter by a centimeter. In comparison, there are 100 billion stars in our galaxy and also 100 billion galaxies in the universe.

The brain contains two hemispheres, left and right, each with four lobes called frontal, parietal, occipital and temporal (Figure 7). The thin surface layer of the lobes is the cerebral cortex, around three or four millimeters deep. It is estimated that the cortex alone contains 20 billion neurons. The cortex is made of gray matter, for it contains the cell bodies of neurons. Deep to the cortex is the white matter, consisting of tracts of axons coming from and going to the cortex, and linking it with other areas of deeper gray matter. Cortical neurons do not function in isolation, but as groups, typically organized in functional columns, running perpendicular to the cortical surface and containing of the order of a million neurons that form a modular structure. Different regions of cortex govern motor control, sensory perception such and sight and hearing as well as sensation from the body surface, and also memory, emotion and speech.

The two hemispheres are connected across the midline by a very large, beam-like structure, the corpus callosum. It is formed of white matter, consisting of some 200 million axons, the axons of specific cortical neurons in each hemisphere that project across to similar regions of the other hemisphere. We shall discuss its function later, but suffice it to say here that the corpus callosum allows one hemisphere to "talk" to the other so that functions that are particularly localized in one side only will be able to influence the other side. This will be of help for fusing visual images from each side of our visual world without

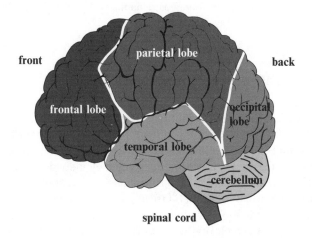

Figure 7 The lobes of the brain

there being a sort of dividing line in the middle, but also for integrating speech and attention.

White matter forms the core of the different parts of the brain, but within this core we again find cellular masses of deep gray matter, such as the thalamus and hypothalamus, and the basal ganglia. The deep gray matter is organized in more or less dense groups, or nuclei. The thalamus is a relay station for incoming sensory information on its way to the cortex, while the hypothalamus helps regulate certain unconscious or autonomic functions, such as temperature control, heart rate and blood pressure. It is also involved with emotions, appetite and thirst, and the release of hormones via the pituitary gland. The hypothalamus also takes part in the important limbic system, vital for emotional behavior and memory (Figure 8). The basal ganglia are very important deep gray nuclei that play a role in motor coordination.

The cerebellum is also responsible for motor coordination and bodily balance. It coordinates sensations, either conscious or unconscious, notably from the balance organs of the inner ear, as well as feedback from the muscles themselves, in order to permit precise control of posture and movement.

The brainstem is at the base of the brain, and joins the rest of the brain to the spinal cord. It is traversed by tracts of white matter going to, for instance, neurons that innervate our muscles, and coming from the cord to, for instance, the thalamus and the cerebellum. It also contains nuclei that play a role in controlling the circulation, respiration, and sleep. In the brainstem there are the small, but important, nuclei of the reticular activating system, that does just that: arousal of the brain from sleep.

Right in the center of all the brain regions is a system of ventricles, containing CSF, which communicate with a narrow space around the whole brain so that it is

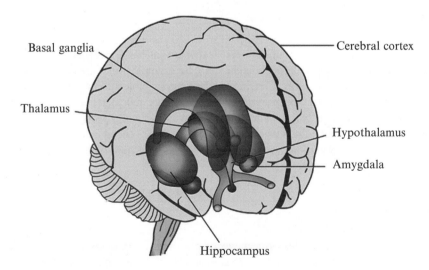

Figure 8 The limbic system

bathed in CSF both inside and out. The CSF protects the brain and spinal cord physically, but it also maintains the environment of the cells and provides for the circulation of certain hormones secreted by the endocrine system. We can sample a patient's CSF by performing a lumbar puncture (also called spinal tap) between the vertebrae of the lower back. In this way we can measure concentrations of, for instance, glucose and electrolytes, to check if they are normal, and also to detect signs of infection of the brain or its surrounding membranes.

The CSF is in a way the liquid, non-electric "humor" of our CNS, but it does not circulate in tubes from heart to brain as the Ancients imagined, and to which they attributed our very being. However, it is true that our good or bad "humor" may depend directly on the flow of "humors" such as hormones.

The Limbic System

The limbic system is not a structure, but a network of nervous pathways involving certain structures mainly situated deep in the temporal lobes, such as the hippocampus and the amygdala. The limbic system participates in the control and expression of emotion, the processing and storage of recent memories, and the control of appetite and emotional reaction to food. All these functions are often affected in psychological diseases. The limbic system is also connected to the autonomic nervous system, that controls such basic bodily functions as heart rate and digestion. Certain psychological problems, such as anxiety, are associated with troubles of the endocrine system and the autonomic nervous system.

Reticular Activating System

Deep in the brainstem are groups of neurons that are called the reticular system. They receive inputs from widely throughout the CNS, for example from the sensory systems, the cerebellum and the cortex. Certain reticular neurons project to the spinal cord and influence cardiovascular and respiratory control. Other reticular neurons project to widespread parts of the brain. Some of them form an ascending pathway, the reticular activating system, that influences our wakefulness and our general consciousness. This system blocks sensory information such as sound and light, and so encourages sleep.

Cortical Areas

Let us come back to the external layer of gray matter, the cerebral cortex. The cortex contains around 20 billion neurons each interconnected by about ten thousand synapses. In all that makes 200 trillion connections. This figure is so astronomical that we can hardly conceive it. If we imagine that each synapse in the equivalent of a letter of the alphabet, then the cortex is the equivalent of a library of 400 million books each containing half a million characters. There is thus a certain sense in saying that the death of a human being represents the destruction of a library. The amazing exapansion of the cortex is a special human characteristic. In reptiles it is minimal; it increases in mammals and reaches a summit of development in our species. If we could spread out the human cortex flat it would have an area approximately equivalent to a square with sides of about 50cm. This is why it is folded, for otherwise it would not fit inside the skull. It is easy to see that the cortex is divided between the two hemispheres, right and left, But it is more difficult to see that each of these is divided into four lobes, frontal at the front, temporal at the lower part of the side, parietal on the side but higher, and occipital at the back.

Certain areas of the cortex are known to be specialized for basic, or primary, functions. Highly oversimplified, we can say that in the posterior part of the frontal lobe is the motor area, anteriorly in the parietal lobe is the somatosensory area, in the posterior part of the occipital lobe is the visual cortex, and superiorly in the temporal lobe is the auditory cortex. They process, respectively, signals controlling motor activity, those dealing with skin sensation such as touch and pain, then vision from the retina and hearing from the inner ear. Between these primary areas are others, occupying a far greater area and volume of cortex in man, that are less easy to delimit but which integrate sensations to permit perception and identification. These are the association areas.

One of the most astonishing regions is that which borders the division between frontal and parietal lobes. In front is the motor cortex, and behind the somatosensory cortex. These areas are organized such that they represent a homunculus with zones containing a map of the half the human body: from top to bottom we find the foot, the leg, the arm, the hand and the face. The area devoted to the hand covers about a third of the surface, thus explaining the extreme sensitivity of our fingers, and their dexterity (Figure 9).

Sensory signals coming from one side of the body toward the cortex, and also those going from the cortex to the muscles, cross in the spinal cord, so that each side of the body is represented in the opposite motor and sensory cortex. Not only that, but one of the hemispheres is dominant: in a right hander the left hemisphere is usually dominant, and the opposite in a left hander. This means that in right handers the left hemisphere contains the speech or language area (Broca's area), as well as reflecting mathematical reasoning, while the right contains artistic sense and intuition.

There have been many attempts to find a way to distinguish the various cortical areas. We know, from what has been said above, and what we shall learn later, that different areas have different functions. But it has been known for a long time that many areas can be recognized by their structure, that is the number and arrangement of the neurons within them. In 1909 the German neurologist, Korbinian Brodmann (1868–1918), described a method for cortical localization based on cellular architecture, and his system has survived the test of time. He gave numbers to

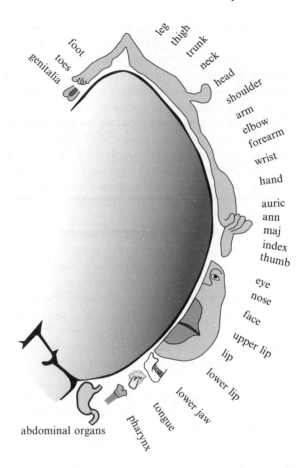

Figure 9 The sensory homunculus. Note the disproportion between the parts of the body depending on their degree of sensitivity

around 50 cortical areas, and we still use these numbers (the "Brodmann areas"), notably in modern functional imaging studies of the brain. For instance, the visual cortex is area 17, the motor cortex area 4 and so on (Figure 10).

The Visual Cortex

Signals from the retina of the eye end up in the occipital lobe. The retina projects in a point-to-point, or retinotopic, fashion through the thalamus, and into the primary visual cortex. A lesion of this visual area will cause partial blindness in the visual field depending on the part of the cortex affected. The image of the visual field is inverted on the cortex, but our brain interprets our visual world in the correct

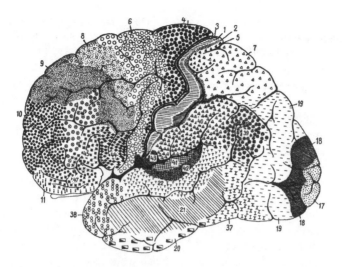

Figure 10 Brodmann's areas of the human cerebral cortex, indicating its main functional zones. He gave numbers to individual cortical areas, and we still use these "Brodmann areas" in modern functional studies. For instance, the visual cortex is area 17, the motor cortex area 4 and so on

sense. If you wear spectacles that invert your visual image, after a few days your brain will discover the trick and you will see again apparently as before.

Anterior to the primary visual cortex there are a number of secondary visual areas that we use to obtain much more information than simply detect light and shapes or patterns. Electrical stimulation of certain of these areas can cause hallucinations involving complex objects or scenes. Partial destruction of some of these higher visual areas can cause agnosia: the subject sees objects but cannot recognize them. He has lost all reference to previous visual experience. This happens with word blindness, for instance: the patient can see the written page perfectly, but cannot decipher a single word, as if the page were covered with unknown signs.

The total destruction of the occipital lobes gives rise to a psychic darkness, much more complete than with a more restricted lesion. Not only is the subject blind but he loses all concept of previous visual experience.

The mechanism is much more complex than people thought initially. We must not imagine that the retina records visual images like a camera or reproduces them like a television screen. Even the retina begins active processing, for it is really part of the brain, as can be seen when one considers its development. As the visual signal progresses from the retina, thorough the thalamus and then into the various visual cortical areas, its components are detected in terms of shape, form, color, depth and movement. Thirty or more specialized visual areas combine their efforts to produce that sumptuous visual experience that we enjoy.

The slightest disorder in any of them can produce aberrant images, such as hallucination or "visions", which capture the imagination of the masses. Individuals may "see" something that is invisible to others and that cannot be photographed.

So mystics see visions and not "apparitions". Our Lady of Lourdes was in the brain of Bernadette Soubirous, and not physically present in the grotto. The phenomenon was real enough, and no trick, but its interpretation by the masses is at fault.

Lesions of limited cortical visual areas can also provoke differentiated loss of sensation. Certain patients suffer from achromatopsia: they see perfectly, but in black and white only. Others have akinetosia: they do not perceive movement. Others still have prosopagnosia: they do not recognize familiar faces. We know indeed that neurons in certain highly advanced visual areas actually "recognize" faces, and especially familiar faces.

This hyperspecialization of the visual cortex helps explain the nature of spatial arts such as painting, sculpture and architecture. It is not necessary to create a visual experience in its entirety, but to present it in a filtered form and to show not what we see in nature anyway, but what we **need** to look at. So we admire the way in which the paleolithic artist in his grotto has been able to suggest the form of a bison by a simple line without cluttering his primitive drawing with excessive detail. Similarly, in prints or photographs in black and white, and in only two dimensions, we discover aspects of reality that color, movement or three-dimensionality may hide. Art is there to excite certain brain areas more than others.

The Auditory Cortex

Just as the retina projects to the visual cortex, the cochlea of the ear projects to the auditory cortex. The pathways for conscious auditory perception abut in the superior temporal gyrus. High pitched sounds are received at the back and low pitched ones at the front of this area, a lesion of which results in total absence of sound perception. Close to this primary auditory area there is a secondary area that helps us recognize sounds. A lesion here, in what we call Wernicke's area, causes word deafness: the patient hears someone speaking but cannot understand, as if the person was using some foreign language. In contrast, the patient can understand the significance of non-verbal sounds.

The Motor Cortex

On the precentral gyrus of the frontal lobe is a sort of keyboard for the contraction of the voluntary muscles of the body. This motor cortex projects to the muscle groups, rather like the homunculus of which we spoke in relation to the sensory cortex, but the size of each part of the body map does not correspond to the muscle mass but to the precision of movement of which the muscles are capable. So the face and hands occupy a large proportion of the motor cortex. Forward of this primary motor cortex we find secondary areas devoted to coordinated motor activity, or motricity, and intention to make a movement oriented toward a precise goal.

A patient with a lesion of certain of these secondary motor areas may not be paralyzed, but his gestures are clumsy and he may lose the ability to perform complex motor tasks learned during his lifetime. Such apraxia in the motor system is comparable to agnosia in the sensory system. Variations on this theme can have astonishing effects. In agraphia, the patient reasons normally, but is incapable of writing, while in anarthria he cannot articulate words to express his thoughts.

One hemisphere is always dominant, similar to what we found in the sensory cortex. As before, this is usually the left hemisphere in right handers, and the right in left handers. This hemispheric dominance plays a role in coordination of symmetrical movement, and may be a good reason not to try to "correct" someone who is left handed.

The Prefrontal Cortex

In that part of the frontal lobe forward of the motor areas we find a region that functions as association cortex for perception coming from other cortical areas. This prefrontal area is particularly well developed in man compared with other primates, more so than sensory cortex such as visual or auditory areas. This was the region damaged in Phineas Gage's accident, and is now known to be particularly affected in certain troubles of perception and behavior, such as schizophrenia.

The Limits of Functional Localization

A relationship between a region of the brain and a given function does not mean that activity of that part causes that function. We cannot exclude that its real purpose is to record that activity in order to inform other regions. We may however be aware that a defect of that region causes loss of the function. The region in question may be essential for a given function, although it may not be localized there as such.

Areal specialization in the brain is a concept that needs careful handing. When we compartmentalize the brain we focus our attention on detail, and may not be able to see the wood for the trees. Different areas have different functions, but they work together, just as a successful sport team relies less on the performance of a single player than on their capacity to collaborate. We shall not learn much about global consciousness by studying an isolated brain region in detail. Consciousness is not localized in a single organ.

The Blood-Brain Barrier

We must open a parenthesis here. The brain may well be the central organ of the body, but it still needs bodily functions to nourish it in the shape of amino acids, glucose and oxygen for example. It receives these via the blood. As we shall see later, an interruption of blood flow to the brain leads to its very rapid degeneration. However, the brain also needs protection against a host of substance that we breathe and ingest that are potentially lethal for neurons. Even certain of our bodily hormones are dangerous. If the blood could feed the brain with anything and everything, the brain would be in chaos.

Nature has therefore provided a barrier between the blood and the brain. Most blood capillaries are lined by endothelial cells through which molecules can pass in order to reach their target organs. But the capillaries of the brain are not organized in quite the same way. The junctions between their endothelial cells are sealed by proteins, rather like mortar between bricks. Small molecules, such as oxygen, but also alcohol, nicotine and heroin, diffuse across these endothelial cells to reach the brain. They manage this because the cell walls contain lipids which attract them. Ecstasy crosses this barrier, but it also damages it for several years. In contrast, the cell walls repel water-soluble molecules, such as glucose, amino acids and vitamins, which are of vital importance to brain function. So nature had to elaborate a transport system for them, using proteins that take up, for example, glucose in the blood, cross the cell, and release it in the brain.

To prevent this mechanism from working too freely the cell wall also contains pumps to reject substances that should not be admitted. Unfortunately this mechanism also prevents certain useful drugs from gaining access, for instance those that fight meningitis or rabies. Later we shall see the difficulties presented by this blood-brain barrier in the treatment of Parkinson disease. Pharmaceutical research is presently working on molecules that can cross the barrier, like Trojan horses, to carry certain drugs.

We shall close this parenthesis by underlining the creativity of evolution in this sophisticated mechanism that protects us and at the same time isolates us. We can neither destroy this barrier, nor ignore it. We must understand it to tame it, according to the general rule that, in order to control nature, we must obey it.

The Development of the Brain

Human development is something extraordinary, governed as it is both by an extremely complex genetic program and by our own individual environment. The combination of these two factors means that we are all different from each other, even identical twins. Many of our bodily features are not specific to man but are common across the animal kingdom. So, how does our brain differ from that of other species that share our planet with us?

Evolution has given us a particularly large brain, with certain special features such as, notably, the fact that a large part of our cerebral cortex is devoted to communication. A major particularity of the human baby is that from around a year of age he will begin to acquire language. When we think of the immensity of such a task we realize that the infant is a master of learning and that he progresses at an amazing speed. Of course his genes have provided structures in his brain that facilitate his apprenticeship, but while his parents admire his progress, his brain is undergoing modifications dependent on his experiences. In the realm of neurons and synapses, new connections are being made, while others are disappearing.

Everything that the baby receives from his parents is coded in the DNA molecule in his cells, which contains the plan of his body. This molecule comprises a few billion bits, that is digital information coded in terms of 1 or 0. Now, an adult brain contains 100 billion neurons, each establishing around 10,000 synaptic connections with other neurons. So the wiring of the brain needs a million times more information than contained in the original cell. Most of the connections, then, must result from the baby's experience as he slowly learns to walk and to talk. A computer comes with all its connections already made. The human brain is also a machine for processing information, but its construction is only completed as it functions. Its builders are the environment in which the baby grows and other humans with whom it interacts.

Even if it may seem contradictory, many neurons will die to leave the field clear for others that are more apt to serve. Acquired experience thus plays an important role in the remodeling of the brain by operating a selection of the most efficient neuronal circuits. A synapse that is used often is reinforced and those that are not used disappear. If two neurons regularly function in harmony, their connection will be reinforced. By using our body and improving our skills we literally remodel our brain. We also remodel our brain by living in a particular environment of sounds and images.

As an illustration, that part of the cerebral cortex representing the ring finger is more developed in violinists, an effect that is more marked the earlier the violin has been learned. A similar effect may be present in relation to the virtuosity of a pianist or singer, the skill of a painter or sculptor, and in general all our learned skills. All apprenticeship corresponds to a particular remodeling of the brain. This does not apply to the same extent to the simple memorization of a recitation by a child or a text by an actor, since they do not correspond to the permanent reinforcement of an ensemble of synapses, although the reinforcement effect is there for just long enough for it to be useful.

Apprenticeship

At birth a baby totally lacks autonomy. He can neither feed himself nor move around alone, and even has difficulty in adapting to the circadian rhythm of day and night. Fortunately for him, a baby is a prodigious learning machine: in a year he learns to walk, at the age of three he can talk fluently, and at seven can read and calculate. But the adventure does not stop there. We are all constantly learning: to use a new electronic recorder, discover the latest video game, perfect our technique at tennis or bridge, or better understand the brain by reading this book.

Since the last century when the Russian physiologist Ivan Pavlov (1849-1936) demonstrated that you can make a dog salivate reflexly at the sound of a bell, scientists have made enormous progress in the understanding of the basis of apprenticeship.

From their studies it appears in particular that the more an individual has access to a complex world, the more his brain develops and more efficient it becomes.

A term which has grown up in the world of neuroscience over the last few decades is neuronal plasticity. We now have evidence at the molecular level that the basis of apprenticeship is to be found in subtle modifications at the synapses.

Language

In most species we observe that individuals communicate with each other by means of cries, signs and odors. But speech remains the prerogative of man. We must define what we mean by speech: human language is a complex system of communication. Indeed, with the help of a finite number of words and grammatical rules, language permits us to express ourselves on any subject of our choice. We do not simply repeat phrases that we have heard: we are capable of inventing others that have never been pronounced and which, nevertheless, have a meaning for those who hear them.

In contrast to reading and writing, language is an innate skill. This means that at birth the baby's brain already possesses certain characteristic structures that will allow him to learn to speak. What is more, children all over the world learning to speak pass through the same stages of development.

The language areas are only situated in the dominant hemisphere, which for most people is the left. As a consequence, if such a person is the victim of a stroke in the left hemisphere he will lose at least part of his capacity for language. By comparing the precise location of such a lesion with the observed loss, we have been able to establish a whole circuit of information processing from thought, through grammatical construction and the search for words, to the production of a phrase.

Thought

"To be or not to be?" That is the question that has obsessed man for millennia, well before Shakespeare attributed the line to Hamlet. To be is to think, but what, in fact, is thought? When we think, we express ideas, most often in the form of words or images. Usually we think consciously, but it suffices to let our mind wander to realize that we are not always master of our thoughts.

How can thoughts suddenly well up from the depths of our brain? For centuries people distinguished between the physical body and the immaterial soul. Where does this philosophical debate around dualism stand today? The evidence is overwhelming that the mind is not an immaterial entity, but rather an emergent property of a hypercomplex cerebral machine. In the world of neuroscience, we have dropped the philosophical term "thought" for the more specific term "cognition", that signifies "to be conscious". Cognitive psychology, backed up by cerebral imagery, today opens up a wide horizon of investigations that will enable us to better understand what the mind really is.

The story of Denis Chatelier, the man with the grafted hands

A French house-painter, Denis Chatelier, lost his hands in an explosion of fireworks in 1996. In January 2000, in Lyon, he received simultaneous grafts of two new hands. This was a world first because such an operation involved mastering several completely different techniques. Firstly, the surgery, by which the continuity between the recipient's arms and the donor's hands had to be established by connecting the blood vessels, nerves and muscles. This meant several specialized teams working for long hours. Then there was the immunological challenge: they had to avoid the rejection of grafted tissue from another body by using appropriate drugs.

But the real unknown factor was neurological. Would Denis Chatelier recover sensation in his new hands to enable him to explore his environment, and would his brain be able to control the movements of his fingers? Even if his new hands were irrigated with blood and stimulated by nerves, would it be possible for him to integrate them in his own body image, already written on his cortex?

After an amputation, the brain notes the loss of a limb. As the part of the motor cortex that once controlled it is no longer in use, it tends to shrink in size and be replaced by neighboring parts. The "hand" area of the cortex is surrounded by those relating to the arms and the face. In the case of Denis Chatelier, cerebral imagery before he received his grafts showed that, indeed, his hand area had regressed, although not completely, and he still suffered from the classic phantom limb syndrome, in which an amputee feels pain from the non-existent limb. He also felt that he could control his lost fingers, although the messages from his brain through his nerves to the muscles could not go anywhere.

During the months following the grafts of his two new hands, Denis's cortex was slowly reorganized to resume its original function. Functional imaging again showed that his hand area was re-expanding whereas his arms area regressed by about a centimeter. After six months he was able to move his new fingers and feel sensation in them. The synapses of the relevant neurons had been reinforced, thus allowing the brain to take over its original bodily function. The changes in his body had precipitated a reorganization of his brain.

Chapter 3
Seeing Through Oneself: Brain Imaging

The present revolution in the neurosciences owes much to the techniques of functional cerebral imaging, which allows us to "see" the parts of the brain being used during specific activities. It represents a considerable step forward compared with the situation faced by early researchers who could only observe a few rare patients suffering from a functional deficit related to a circumscribed lesion, either visible at clinical examination or discovered *a posteriori* at autopsy.

Certainly, all the classic techniques of imaging can be, and have been, applied to the brain, although some of the more dangerous ones are now no longer necessary. For example, angiography consists of injecting a contrast medium into the cerebral blood vessels and then taking a radiograph (X-ray). This enables the radiologist to reveal anomalies of the vascular system. However, modern cerebral imaging allows us to do two things, in the healthy or diseased brain: we can reveal the normal anatomy of the brain, and we can examine its function. The pictorial quality of the results depends on whether we need a high-resolution spatial image or rather an examination as a function of time.

Magnetoencephalograhy (MEG)

This technique uses electrodes placed on the skull to measure the very weak magnetic fields generated by electrical activity in cortical neuronal circuits. In practice, in order to detect anything worthwhile, we need 100,000 to a million neurons to be excited simultaneously, which corresponds to the functional cortical columns that we referred to earlier, that is cylinders of about 3mm in diameter through the thickness of the cortex, which is also about 3mm.

MEG was developed between 1986 and 1972 by David Cohen at the Massachusetts Institute of Technology (MIT). Since then it has been much improved thanks to the availability of more and more powerful computers, for it is not enough to simply measure the magnetic fields: the currents that are responsible for them must be localized. The main drawback of MEG is the difficult and onerous nature of the calculations that necessitate considerable computational facilities. A strong point of MEG is its excellent response in real time, that can be a fraction of a millisecond.

Only electrodes actually in the cortex can do better, but that is a potentially dangerous invasive technique. In comparison, other imaging techniques that we shall
discuss later have a time function of anything from seconds to minutes.

MEG is often associated with EEG (electroencephalography), which measures
variations in voltage due to cortical electrical activity, again through electrodes
placed on the scalp. EEG is a much older technique, described by German psychologist Hans Berger (1873–1941) in 1924. A few years ago, EEG was studied via
16 electrodes. Now, high spatial resolution EEG uses 200 or more electrodes. The
time function is similar to that of MEG, but the latter is spatially more precise
because currents in the scalp vary with the resistance of local tissues. Like MEG,
EEG is also non-invasive, but has the advantage of being relatively cheap and not
requiring specialized medical personnel.

In MEG the magnetic field is measured in a classic manner, by observing the
current induced in a coil. Obviously the variation of the field due to activity in a
single neuron would be undetectable. So we can only observe function of neuronal
assemblies or columns. Even so, the fields that are being measured are so weak that
they can be masked by electrical parasites. As a comparison, these fields are a thousand
times weaker than those produced by the patient's heart, 100,000 times weaker than
those produced by a car passing 50 meters away, and a billion times weaker than the
earth's magnetic field. This means that the equipment must be used inside a heavy
metallic shielding that will prevent outside parasites from spoiling the measurements. In order to record the feeble currents induced by these fields, special detectors
are used, cooled by liquid helium to produce an effect of superconduction, that is
with no electrical resistance.

MEG is a good example of the sort of measuring system available to modern
science through the combination of various sorts of technology. We must dispose of
an enormous range of resources in order to overcome problems that are simple to
formulate: we want to measure, at a distance, the weak currents produced inside the
brain by measuring their associated magnetic fields.

Tomography: CT and PET

CT (computerized tomography) and PET (positron emission tomography) are related.
The former provides good spatial precision, while the latter gives information about
the function of an organ. Both can help detect the presence of tumors, and both use
radiation, although of different natures. The modern CT scanner still uses X-rays
(electromagnetic radiation similar to radio waves but at a much higher energy).
Wilhelm Conrad Roentgen (1845–1923) discovered X-rays at the end of the nineteenth century, for which he won a Nobel Prize in 1901. They are used classically
to provide two-dimensional images of the body. The absorption of X-rays varies
according to the nature of the tissues penetrated (bone has a high absorption and
appears white on the resultant radiograph, whereas soft tissue is less absorbent and
appears black). So, one obtains an image of internal organs, albeit at low contrast,
and with no functional information.

Tomography refers to the study of "slices" of the body. It originally used classic radiographic techniques, but where the X-ray source swung back and forth at an angle so that the object of interest, for example a tumor deep in a body organ, remained relatively sharp, whereas surrounding tissue was blurred in the moving beam. The results were often so blurred as to be difficult to interpret.

A considerable refinement, the CT scanner uses an X-ray source that rotates around the body of the patient, through $360°$. Many two-dimensional images are captured as the X-rays penetrate at a series of angles, so that sections of the body can be recreated mathematically by the reconstruction of the volume of an object from plane projections. This technique was developed in the 1960s and rewarded in 1979 by a Nobel Prize for Allan McLeod Cormack, a mathematician, and Godfrey Newbold Hounsfield, a radiologist.

PET scanning is also based on the use of radiation, not from a source outside the body but from radioactive tracers injected in the patient. Naturally, the tracers are selected for their rapid disintegration so as not to irradiate the patient for too long, but long enough to make measurements that are reproducible. For example, we have radioactive carbon 11, that emits only beta radiation, that is to say positrons, analogous to electrons except that their charge is positive rather than negative. These particles penetrate a few millimeters into the body before encountering an electron. When the two combine the electrical charge is cancelled and two photons result, like flashes of light. These separate in opposite directions and are detected simultaneously by two instruments placed at opposite extremities of the detector system. The precision of the resultant spatial localization is limited by the path of the positron before it releases the two photons and by the fact that the photons are not released absolutely diametrically opposite each other.

The aim of the CT scanner is to obtain a clearer image than we can obtain from a PET scan. The former gives a good spatial anatomical image, the latter a functional image. The main use of PET is the detection of tumors. For this the sugar molecule desoxyfluoroglucose marked with radioactive fluorine is used. Tumor cells utilize more glucose than healthy cells, and a PET scan using this tracer enables the detection of very small tumors. In another use, we can accurately analyze the amount of dopamine present in the brain of a patient with Parkinson disease.

Magnetic Resonance Imaging (MRI)

MRI was originally developed as an anatomical imaging method, but today it is used to obtain functional images also. Its spatial accuracy is excellent, of the order of a millimeter. It is based on a principle described in 1946 independently by two physicists, the Swiss Felix Bloch (1905–1983) and the American Edward Purcell (1912–1997). Its underlying physical methodology is not as obvious as in the two previous techniques, that basically measure a magnetic field or electromagnetic radiation.

For an MRI the patient is placed in a uniform magnetic field that serves to orient in the same direction the magnetization of hydrogen atoms, which are numerous in the human body, 80% of the weight of which is composed of water. An electromagnetic impulse of appropriate frequency upsets this orientation for a brief instant. As the magnetic fields return to their initial position a current of the same frequency, or resonance, is induced in a detector coil that allows the level of the magnetization to be measured. The resonance frequency is proportional to the magnetic field used. By creating a magnetic field that is variable in space, it is possible to spatially localize the different parts of the body because they emit different frequencies. We can measure not only the position of organs, but also their speed or acceleration, for example that of blood in vessels. So MRI can be used for functional studies (functional MRI or f MRI) in addition to anatomical ones.

The advantage of fMRI over PET is obvious, in that we do not need to inject radioactive substances. For studying the brain we use hemoglobin as a tracer, for its magnetic properties differ according to whether it is transporting oxygen or not. The contrast of the images depends on the oxygen level in the blood. When neurons are active they consume more energy and the blood vessels need to bring them more oxygen. So the MRI signal is greater in the active zones of the brain and can be visualized. As for the techniques described earlier, the MRI signal is weak and may be drowned in the background noise, so it is necessary to extract the signal by statistical processing by a computer.

Thanks to MRI researchers are able to study more and more complex brain functions such as pain, emotion and calculation. It has become an indispensable tool for functional brain research. A surgeon, before removing a tumor, can undertake a range of motor, sensory and cognitive tests before the operation in order to avoid damage to vital areas during surgery.

Research Using Imagery

As we have already mentioned, medical imagery raises a number of distinct problems. We can define neurological phenomena that depend on different brain regions and may help us localize those regions or understand their function, but we still need to measure their characteristics in very difficult circumstances and then reconstruct what happens inside the skull using measurements made outside.

All the various methods of imaging provide us with rich and varied information on the human brain. Image analysis strives to extract the maximum information for us to make a diagnosis with a view to a therapeutic intervention. In the domain of brain imaging we have reached a stage of transition: we are switching from the use of the image by a medical practitioner simply to see what is happening purely qualitatively, to use of the image in quantitative terms. Engineers are striving to provide tools to allow precise quantitative measurements by exploiting the totality of the recorded information from all imaging sources, whether anatomical or functional, whether derived from EEG or MEG, CT or MRI, or PET. Each method has its strong and weak points. The problem is to combine these images in order to arrive at a final diagnosis concerning the only thing of real interest, the brain itself. CT and MRI give precise information on anatomy, just as we see when we dissect the brain. PET only produces a poor anatomical image, but gives an excellent functional image. Combination of the two sets of information gives us precise results concerning the zones that are active, how active they are, and where they are.

Such work had a direct clinical application. Neurosurgeons uses vast numbers of brain images for diagnostic purposes and, more recently, in the preparation of operations, such as for the treatment of Parkinson disease, in which they need to implant electrodes at precise coordinates in the brain in order to suppress the disordered body movements. Also, in the context of current research into the prevention of this disease, it is necessary to inject retrovirus into the substantia nigra with a precision of a millimeter or so.

The Brain as a Black Box

The aim of the laboratory of Touradj Ebrahimi at the EPFL is to investigate pathways for access to the brain other than through the primary senses of touch, vision, hearing or smell. For example, it would be marvelous for patients who have lost the gift of speech to be able to dictate texts directly by brain power, or for the paralyzed to be able to control their environment, such as opening a door or a window, in a similar way. To do this, Ebrahimi studies the brain as a black box. He excites the brain and measures the results. The simplest way to do this in an engineering laboratory is to use the EEG. This avoids the medical supervision that would be necessary for the injection of tracers, as well as the availability of extremely costly machinery. Indeed, the latter are unnecessary as the object is not to produce high resolution anatomical or functional images.

From the recorded EEG signals he attempts to create an abstract model of the brain, an ensemble of equations that relate signal and response, without being concerned about the organization of the brain itself, neither its internal structure, nor the partition of function between areas, nor the interaction of these areas among themselves. His model is therefore purely theoretical and in no way a spatial representation of the real brain. He sees the brain as a black box, represented in completely abstract n-dimensional space, unrelated to biological localization of processing. With such an approach it is possible to see a correspondence between a brain function, such as a movement of the right arm, and a portion of abstract space.

Admittedly, the classification of actions in this space is far from perfect. But Ebrahimi has discovered that the human subject is capable of correcting the function of his brain to adapt to the mathematical model to which he is reacting. The subject manages to modify the signals he emits to better enter into the abstract classification. In other words he learns to use this technique just as he learns to walk or ride a bicycle. This approach will doubtless be of relevance to learning, for learning a language or mathematics is an operation that takes place essentially by the reinforcement of specific synapses in the brain.

Ebrahimi's method is related to another, but one that relies on a much more consequent technical arsenal. Since 2003, scientists led by John Donoghue of Brown University have succeeded in implanting in the motor cortex of tetraplegics, or other patients with motor handicaps, a square module of 4×4 mm, containing a hundred electrodes that capture signals emitted by his brain. After a period of apprenticeship patients are able to control a prosthesis of the arm and the hand that enables them to grasp objects purely by signals from their brain. This method, called BrainGate, is more efficient than the use of the EEG because the signals are derive directly from the brain and have not been interfered with or attenuated by the scalp. On the other hand, this technique has the disadvantage of presenting a permanent connection between the interior of the skull and the outside world, with the ever present risk of infection.

Virtual Reality

Olaf Blanke at the Brain Mind Institute of the EPFL has linked an EEG apparatus to a computer screen, which is the subject's only universe. It is far removed from reality, but gradually he integrates with virtual reality. In the first stage the screen simply replaces the natural visual field. In a second stage the system also allows for the integration of the subject in the visual presentation. Using his own brain, of which the activity is monitored by EEG, if the subject decides to move to the right, the program modifies his animated representation on the screen. In the third stage the whole of the brain's activity will be integrated in virtual reality. The laboratory also uses fMRI, which allows a better penetration into the spatial structure of the brain than high resolution EEG, with a precision of about a millimeter.

With these same research tools the laboratory aims to refine diagnosis in the context of a clinical neurological service. It will target patients who have problems of movement perception, whether visual or auditory, some of whom cannot even detect direction of movement or recognize a face. For this, we must try to understand the normal and pathological mechanisms of how we perceive movement, either using vision or hearing, and how these two senses interact. For instance, if we see a fly going from left to right, whereas the buzzing goes from right to left, how do these contradictory sets of information interact? How does the brain perceive its own body? What brain structures make us conscious of our body? For indeed, just like the outside world, our body image must be constructed and perceived.

When surgeons are forced as a last resort to resect epileptic cortex, it is important to localize speech and motor control. Today they have to stimulate with electrodes in the brain itself in order to localize the relevant areas. In a few years this invasive process should be replaced by imagery that will allow even more well-targeted surgery.

The story of the man who mistook his wife for a hat

In his career as a neurologist in New York, Oliver Sacks met several patients whose life had lost all coherence, but none struck him more forcibly than the one who inspired the title of this story. Indeed it became an immensely successful play directed by Peter Brook.

At the end of a consultation, the patient literally tried to take his wife's head in order to put it on his head, without the remotest vindictive intent! Indeed his wife considered his gesture as quite routine. She had learned that her husband no longer recognized her, for he suffered from agnosia, a selective loss of his capacity to recognize objects, and particularly faces, for their true value. His vision in itself was excellent, but his selectivity for objects was lost. He was a professional music teacher and was in no way hindered in the exercise of his profession.

Sacks reports a discussion between the patient and himself about a glove. The patient recognized full well that the object was made of leather, of a certain color, and consisted of five fingers, but could not relate it to being a glove. After considered reflection, he decided it was a special purse designed to keep five different sorts of coins. Similarly he did not recognize the faces of famous people but was able to identify them through minute details, for example Winston Churchill by his cigar. A minute part of his occipital lobe was damaged and this produced this strange alteration of his personality.

Chapter 4
Dispersed Memories

We do not have a memory: we have several memories. There is the memory that we use, without noticing, to allow us to live from day to day, and also different forms of memory that situate us in our own time frame, and of which we are conscious.

Our unconscious procedural memory is one of the most important. It is here that we store the actions that we need to accomplish any task, like walking, riding a bicycle or eating with a fork. Procedural memory is built up by the repetition of the same actions that end by being imprinted in us without our being aware of them, without our having to think about them. It is even difficult for us to describe the details of these actions.

Another form of memory is conscious: it helps us recall more-or-less precisely dated and localized events, and we call it episodic memory. Its establishment is in several stages. First of all our sense organs, for touch, vision, hearing, olfaction and so on, feed our sensory memory, that is of very short duration, a few seconds. We can take pleasure in testing this when we analyze the impression left after sampling a good wine: if we estimate the time, the "codalie", that the taste remains on our palate, we note that it is rarely more than about ten seconds. These sensations are then stored in our short-term or working memory that lasts longer, a minute or so. This is the memory we use between reading a telephone number and dialing it. For most people the limit is about seven numbers. Long-term storage in episodic memory only happens if these temporary memories are reinforced by repetition and then encoded during sleep.

Of a similar nature to episodic memory is semantic memory in which we store not memories of our life experiences but general knowledge, like questions that would be worthy of a television quiz, such as the date of the Declaration of Independence or the Battle of Hastings, or how many symphonies Mozart composed.

Visual, auditory and conceptual information can be followed in real circuits in the brain. The first stage is coding of information from the sensory organs. The left frontal cortex and the hippocampus are involved in the process of its transformation into a storable form. This coding depends on the level of our attention and our interest in the information in question. The important events in our life are preserved, while most of the rest disappears for ever.

The second stage is permanent storage in the cerebral cortex, for example in occipital cortex for visual memories and temporal cortex for semantic memories.

The third stage is consolidation. To avoid being forgotten, information must be continuously consolidated, in some cases over a period of years. The hippocampus has a pivotal role as part of a cerebral circuit belonging to the limbic system, the Papez circuit, named after American neuroanatomist James Papez (1883–1958) who first described it in 1937. In addition to the hippocampus it includes the fornix, the mamillary bodies of the hypothalamus, part of the thalamus and the cingulate gyrus.

The fourth stage is recuperation. When we recall an event, its different elements are reconstituted. This task is accomplished by the hippocampus for automatic recall and by the right frontal cortex if we need to make a mental effort to remember.

Storage in Neurons

We are familiar with physical memory outside our own bodies. A sheet of paper covered with writing, an old computer disk coated with a million minute magnets, a compact disk with a spiral of tiny bumps and pits. This type of memory is localized, in that a given bit of information is located at a specific place: a given word is on a given line, a given passage of a Mozart symphony is in a specific part of the CD.

Our memory is not like that. There is no molecule located precisely in my brain that stores the taste of the wine that I recall, the smile of a loved one, the voice that I recognize among thousands. How **is** memory stored in our brain? The beginnings of a reply were suggested by Cajal in 1894. Learning facilitates a swelling and growth of the synapses that interconnect neurons. The electrical activity that represents a certain experience in an assembly of neurons persists beyond the precise moment of that experience. This causes modification of the synapses so that those that are active are reinforced.

This simple idea, formalized by Canadian psychologist Donald Hebb (1904–1985) in 1949, today still forms the basis of our understanding of memory. It is easy to conceive that memories or skills that are elicited regularly will be more deeply imprinted than an event of little significance about which we think for an instant and then no more. Important events, whether happy or sad, whether they elicit pleasure or anguish, about which we think constantly during subsequent days will remain in our memory indelibly. There is only one way to learn a text, and that is to repeat it over and over. On the contrary, a poem learned in infancy, and that we have never rehearsed since, finishes by being deleted from our memory apart from disjointed phrases.

For if cerebral activity creates or reinforces synapses, inactivity removes or weakens them. The reinforcement of certain sets of synapses is inevitably accompanied by weakening of others, for memory does not have an unlimited capacity. In our view of the brain as we understand it, most life events have never been committed to long-term memory and, even among those that have, most will be lost subsequently. According to dualism, the immaterial spirit would recall all it had experienced, although some thoughts might remain buried in the subconscious. The subconscious, or unconscious, so dear to the psychoanalysts, would correspond to thoughts that are stored but temporarily inaccessible due to certain inhibitions.

Sleep plays a special role in the fixation of long-term memory. If you learn a lesson in the evening it is better retained the next morning after a good night's sleep than the same lesson learned in the morning and tested in the afternoon. In contrast, too short or too agitated a night disturbs imprinting on memory. There is in fact a difference between the types of memory. Procedural memory, that of actions, is fixed in the brain during episodes of paradoxical or REM ("rapid eye movement") sleep, when the sleeper moves his eyes and dreams. Episodic or semantic memory is consolidated during periods of the other type of deep, slow-wave sleep. This consolidation of memory during sleep suggests that the sleeping brain may replay the neuronal activity that it has experienced during the previous day, and, in this way, reinforce the already excited synapses.

The Generation of Neurons in the Canary

Could new neurons be created to store new information? Not long ago we did not even consider such a possibility. Today the answer is less clear-cut. We now know that neurons can still be generated in the hippocampus in young adults and then migrate to other brain regions. It has also been shown that apprenticeship improves the survival of these new neurons. This is obviously an area that is well worth exploring further in the hope that we may be able to compensate for the loss of neurons that is the underlying cause of many neurological diseases, as we shall discuss later.

The story of the discovery of this mechanism of generation of new neurons in the adult is worth recounting, for it is a good illustration of how research works. At the beginning of the 1970s the dogma of a slow but irreversible loss of neurons in the adult was firmly established. However, a scientist at Rockefeller University, Fernando Nottebohm, was intrigued by the song of the canary. This song plays an essential role in the survival of the species because the male uses it to attract the female and to chase other males from his territory. It is not simply a behavioral trait designed to delight our ears and brighten our dreams.

The canary's song is not programmed in his brain at birth. He learns by imitating the song of other males, and this is in itself variable from one individual to another. The canary learns to sing as we learn to talk. He only sings in the spring, the mating season, and his song varies from one year to the next. How? Is it memorized in the synapses of certain neurons rather as we have seen elsewhere?

To the surprise of Nottebohm's team, they discovered that the canary produces new neurons continuously that take their place in the song center to replace old neurons that die when the bird ends its singing season. The survival of the neurons and their capacity to learn a new song depends on the level of testosterone, the male hormone, that is itself regulated by the length of daylight.

The story of the canary that is able to renew its neurons remained a biological curiosity for a decade. However, already in 1965 Joseph Altman and Gopal Das at the Massachusetts Institute of Technology (MIT) had suggested that there could be neurogenesis in adult rodents. In the 1980s and 1990s it became clearer, notably through the work of Michael Kaplan, Elizabeth Gould and Fred Gage and their colleagues, that indeed certain parts of the mammalian brain, such as the olfactory system, but also the hippocampus, could still produce new neurons during adulthood. This process even extended to the primate, and human, brain, although others contested this. The creation of neurons seemed better when the animal was exposed to frequent stimulation, for instance by odors to excite the olfactory system. As it stands at the moment, we have to conclude that the creation of new neurons in the adult may well be another mechanism for apprenticeship.

The Hippocampus of the London Taxi Driver

Research has demonstrated the central role of the hippocampus in memory. To illustrate the nature of memory for places that enables us to find our way around a familiar town, in 1997 Eleanor Maguire and her colleagues at University College London studied the brains of London taxi drivers with MRI. These rather special subjects are required to follow intensive training for two years at the end of which they are expected to have acquired a detailed knowledge of their particularly sprawling and complicated capital city. A tough examination serves to eliminate all but the best trained drivers with regard to London traffic.

This study established that the posterior part of the hippocampus of these successful drivers is more developed in MRI scans than that of a set of controls. The conclusion was that this part of the hippocampus plays an important role in the acquisition of spatial memory. The longer the drivers' training, the more this brain structure was reinforced. In other words the hippocampus remains relatively naïve before training, but grows in efficacy during training and due to training. This study also demonstrated that the hippocampus of the subjects was indeed active during spatial memory tasks.

Further to these results, in 2006 Maguire studied a London taxi driver who had hippocampal damage: although he could still find his way round the main London streets, he got lost when he steered off the primary roads.

Two Diseases of Memory

The commonest dysfunction of memory is amnesia, or loss of memory, but it is worth noting that there exists an opposite phenomenon, hypermnesia, either total or partial. As an example of a special form of hypermnesia, Mozart had the capacity to recall music that he had heard only once, and be able to replay it by heart. At the age of 14 he transcribed the score of Allegri's *Miserere* that he had heard for the first time the same day. The great conductor Arturo Toscanini knew 17 operas by heart. If such phenomenal memory constitutes an advantage in certain fields, it can be a real handicap in some people who remember everything, although they may suffer from other mental deficiencies.

The story of Kim Peek, the man who knows too much

Kim Peek was born in Salt Lake City in 1951 and he soon struck his entourage by his outrageous memory. Since the age of 18 months he could recall any book read to him. Later, he could read a page of text in a few seconds, which means that an average novel would take him an hour and a half. He now knows by heart a library of over 9000 works from authors of classical literature to musical scores. He is interested in history, sport, cinema and geography. He piles up in his limitless memory all the data he ever encounters. He was the inspiration for the "savant" character of Raymond in the film "Rain Man" in 1988.

Cerebral imaging revealed an abnormally voluminous brain, well into the highest percentiles of the population. But his brain showed several malformations. The corpus callosum, the giant bundle of nerve fibers that connects the two cerebral hemispheres, was totally absent and his cerebellum was malformed. The absence of a link between the two hemispheres is considered a possible explanation of his incredible memory. Normally, the dominant hemisphere, the left for most right handed people, can inhibit the right. If this control does not exist it is possible that the right hemisphere might develop abnormal capacities. The anomalies of his cerebellum should be manifest as reduced motor skills: indeed Kim has an unsteady gait and cannot button his clothes.

Apart from these minor handicaps Kim's intelligence is normal, although rather unequal in various domains. Not only can he memorize a host of facts, he can interrelate them. When he speaks of a composer, he evokes not only the particularities of the score he has memorized in detail, but he can relate the music to the historical context of its composition. He learned to play the piano and, like Mozart, possesses a stupefying capacity to learn a score heard only once, even years before. So we may well wonder if Mozart's astonishing gifts might, too, have been related to abnormalities of his brain. It is impossible to answer that question, for the composer's body was buried in a pauper's grave.

Chapter 5
The Prevention of Parkinson Disease

Parkinsonism is a general term to describe symptoms caused by degeneration of part of the brain devoted to control of bodily movements. Under this overall heading we include not only Parkinson disease (PD) in its strict sense, but also analogous conditions, such as the tremor that affects many older people who do not have PD itself.

PD is caused by the loss of neurons in a circuit that controls movement and that normally produce dopamine, a neurotransmitter that propagates the nerve impulse across the synapse. It is not difficult to see that such a loss will disturb neurotransmission and thus the control of movements, giving rise to tremor, stiffness of the limbs, slowness of movements and even the loss of the ability to move, that are the best known characteristics of the disease. Initially PD is different from Alzheimer disease in that it attacks motor function rather than cognition, but ultimately many PD patients suffer from a psychological degradation.

Parkinsonism is due in about 80% of cases to PD itself. Its name is derived from that of James Parkinson (1755–1824), an English physician who first described it in 1817. His description concentrated on muscular rigidity and spasms associated with tremor. In 1860 French neurologist Jean-Martin Charcot (1825–1893) proposed to name the disease after Parkinson. It is estimated to affect between 6 and 16 million people in the world, and is likely to double by 2040. In most countries it affects between 1 to 3% of the population. Symptoms usually appear in people over 50, but in 5% of cases it is diagnosed before the age of 40. Men are affected slightly more frequently than women.

Several clinical signs and symptoms exist. There is often tremor at rest which disappears when the patient makes a voluntary movement. Commonly there is increase of muscle tone leading to rigidity. Another feature is slowing of movements (bradykinesia). In severe cases there is absence of movement (akinesia), often associated with a loss of facial expression. The patient tends to lose his sense of balance, and falls are frequent. His gait becomes a mere shuffle, and speech is slurred. Among the early signs of PD there can be the rather curious micrographia, in which the patient spontaneously reduces the size of the letters as he writes a line of text.

PD is due to pathology of the brainstem, specifically that part of the midbrain called the substantia nigra. If we dissect that region, we see a small elongated nucleus with a black pigment, containing neurons that release the neurotransmitter dopamine. This is the main, and perhaps first, pathway that suffers in PD, but other dopaminergic pathways can be affected leading to the other symptoms of the disease over and above the purely motor ones. At birth we possess about a million dopaminergic neurons in our substantia nigra. Symptoms of PD appear if we lose 60 to 70% of them. It is amazing that loss of a few hundred thousand neurons among the billions that our brain contains can have such a devastating effect.

Even if PD is not a fatal disease in itself, it represents a painful ordeal for the patient and his entourage because in the present state of our knowledge it is impossible to prevent it, to cure it or even stop its progression. The best we can do is to attenuate its effects as they progressively destroy the patient's quality of life, with uncontrollable tremor of his limbs, impossibility to walk, digestive problems, and a deterioration of affect and reasoning. The patient is confronted with an inexorable fate: a gradual loss of control of his own body, with no hope of stopping the process. The therapeutic measures that exist cause so many side effects that other medication is needed to counter them. The patient's day becomes a depressive marathon of pill taking. Recent work, still in progress, raises the hope of some solutions, if not for a cure at least for prevention. Let us try to establish the present state of affairs.

A Problematical Treatment

In the present state of our knowledge it is impossible to stop the progression of PD once it has produced clinical signs, for this would imply reversing the process of degeneration of neurons. However, ongoing research that we shall discuss later points to some hope for the not too distant future.

Even if the treatment of PD has changed in the last few years, it still depends, for the most part, on the prescription of levodopa (L-dopa). This molecule has a number of side effects which have to be treated by medication or even surgery. In addition it is necessary to envisage treatment of the non-motor complications of PD. Last but not least, we must not forget physical, psychological and social support for the victims of PD as well as their families.

Although research into improving the efficacy and limiting the side effects of chronic medical treatment of PD has met with some success in recent years, the riddle of the ideal treatment to stop or reverse the degenerative process remains unsolved. A recent study demonstrated no convincing proof that any form of present-day treatment had any protective benefit for neurons. That is why the objectives of therapy can be summarized as limiting the impact of the motor symptoms of PD and avoiding the complications of treatment. However, in general the principle objective of treatment is to preserve as long as possible the quality of life of Parkinsonian patients. To attain these objectives certain decisions must be weighed in the balance. At what stage should active treatment for a patient begin? What therapy should be adopted? Should there be recourse to radical procedures such as surgery, and when? A priority is to gain the patient's confidence and establish an active relationship with him. To educate the patient is essential and the doctor's role is to provide necessary information and direct the patient to the relevant resources.

When to begin active therapy is one of the most critical decisions in the treatment of PD. The decision to start or withhold treatment depends on the circumstances in which the patient finds himself, including his age and professional situation. There is no proof that early treatment, with whatever drugs, slows the progress of PD, or that treatment is beneficial before the symptoms have had an important impact on the patient's quality of life. In the absence of proof to the contrary it remains preferable not to treat PD too early, but to follow its evolution regularly. In spite of everything, PD will finish by interfering with the normal functioning of the body and symptomatic treatment will become necessary. It must be clear that the aim will be to palliate the effects of the disease, not to cure it.

L-dopa is the bastion of the treatment of PD. It is inactive in itself but after crossing the blood-brain barrier it is converted to dopamine, the missing neurotransmitter, by an enzyme, dopa-decarboxylase. Curiously, this production of dopamine can be counter-productive, for much of the effect of the "new" dopamine is on systems outside the ones we wish to treat, causing undesirable side-effects. To reduce these, L-dopa is commonly administered with an inhibitor of dopa-decarboxylase. Unfortunately L-dopa is degraded to a large extent during its absorption, so reducing the quantity available to the brain. Further, it is eliminated rapidly, in 60 to 90 minutes, thus compromising a relatively constant supply such as is needed by the brain. Three approaches have been tried to attenuate these fluctuations of L-dopa: suppression of metabolic pathways, utilization of a controlled release form of the drug, and the fractionated use of multiple small doses. This last approach can result in insufficient drug being administered if the total daily dose is not increased.

Even if L-dopa remains the most efficient way to treat the motor effects of PD, all patients risk complications in the long term. Thus those with early onset of the disease, who are especially at risk of fluctuations in their response (most, if not all, within five to ten years), must begin to take a dopamine agonist. An agonist acts by direct stimulation of dopamine receptors, thus avoiding the need for chemical conversion of L-dopa and its storage. After L-dopa itself, these agonists form the most effective therapy of PD and may be preferred as an initial treatment in most cases. Indeed, many consider that treatment of young patients should not begin with L-dopa.

In 2000, results of studies on a dopamine agonist, ropinirole, provided evidence in their favor as medication for early-stage PD. In a comparative study of two groups of young patients treated with on the one hand L-dopa and on the other with ropinirole, those treated with the agonist showed less problems of motricity than those using L-dopa. Scores for achievement of everyday activities were similar in both groups. Such studies demonstrate that therapy with dopamine agonists during the early days of PD can be practical, safe and effective, and they might be envisaged as a treatment of choice. However, only about 50% of patients finished their trial period of five years, and some two thirds of the ropinirole patients also needed supplementary L-dopa. Side-effects of dopamine agonists are also reported, so they are not miracle drugs.

The advantage of dopamine agonists may be related to their longer persistence. Their period of activity is three to five times longer than that of L-dopa. They have also proved their value as supplementary treatment in the final stages of PD by reducing the dose of L-dopa needed and increasing the activity of L-dopa, leading to reduced tremor. We do not yet know whether a combined treatment with L-dopa and a dopamine antagonist might further reduce the onset of complications of therapy.

Finally, we should note that apomorphine administered by subcutaneous injection or by perfusion pump can be useful, especially in the terminal stages of PD to treat the appearance of sudden bouts of tremor. However, it is not recommended in the early stages of the disease.

The Complications of Therapy

Because of the inexorable progression of PD, complications are inevitable. Various factors can provoke fluctuations in the body's reactions to L-dopa. These include the environment in the digestive tract, including stomach acidity. Variations in brain metabolism can also affect the pharmacodynamics of L-dopa. These can include reduction of dopamine stores, damage to dopamine-producing neurons by dopamine derivatives, or changes in dopamine receptors.

The commonest fluctuation is loss of efficacy characterized by a reduction in the duration of its effective period after successive doses of L-dopa. This loss of efficacy is common and predictable as a result of diminishing function of the target cells. With time and the progression of the neuronal degeneration these fluctuations due to a predictable relative loss of efficacy can be succeeded by a violent and unpredictable "on/off" phenomenon. This is characterized by a sudden transition from a state in which the symptoms are under control ("on") to a period where they are out of control ("off"). During this phase patients can present with sensory or psychiatric symptoms. They may complain of pain, sweating, constipation or breathlessness, most marked during "off" periods. These "on/off" swings are associated with a lowered threshold for dyskinesia. The handling of motor fluctuations can imply a delicate equilibrium of the drug regime in terms of dosage and drugs used. The aim must be to obtain a maximum effect without exceeding the required dose.

Orthostatic hypotension (dizziness when rising to or maintaining an upright posture) is frequent in PD. It is either provoked by the disease or its treatment. Although most drugs used in PD may cause or worsen the problem, dopamine agonists are the most likely culprits in orthostatic hypotension. If the hypotension is minor, it can be managed by taking care to avoid suddenly rising from a reclining to a standing posture, and ensuring that some form of walking support is always available. If the problem is severe, drugs to raise blood pressure can be envisaged, but in elderly people they can produce secondary effects such as edema, congestive heart failure or hypertension at rest.

Constipation, due to reduced intestinal motility, is a frequent complaint in Parkinsonians. It tends to be exacerbated by most of the drugs used, as well as by a fiber-poor diet, lack of adequate liquid intake, or lack of exercise. Attention to these factors can resolve the problem. Fiber-rich food supplements, products that soften the feces or, in serious cases, enemas or drugs to quicken intestinal transit can be envisaged. Nausea is another undesirable effect of many drugs. It can be reduced if L-dopa is taken with food, but this solution can also provoke deficiencies in drug uptake because absorption may be reduced.

Dyskinesia can be manifest by unwanted involuntary movements or by difficulty with voluntary movements. This can typically be due to the disease itself or its inadequate treatment. Dyskinesia is usually associated with peaks in the kinetic profile of L-dopa, but it can happen between "on" and "off" phases of an unstable regime. Dyskinesia can be improved by slight reduction in the dose of L-dopa and the addition of a dopamine agonist to help in motor control. Diphasic dyskinesia can be treated by imposing very short intervals between drug doses which, in some cases, can only be obtained by using a liquid preparation of L-dopa or perfusion with L-dopa or a dopamine antagonist. Surgery to lesion the globus pallidus or subthalamic nucleus may be necessary, as we shall discuss below.

A typical form of akinesia, found mainly during "off" phases, but also less frequently during an "on" phase, consists of a hesitation or "freezing" when the patient tries to begin to walk, or it can take the form of a sudden immobility while he is walking or performing a task. This effect is exacerbated by stress, and its cause is unknown.

Postural instability is a common characteristic of advanced PD and does not respond to L-dopa. Indeed, L-dopa may just be sufficient to allow a patient to begin to walk, thus exposing him to the danger of falling. There is no effective treatment for instability: ambulatory patients in whom it is a known risk should always be provided with a walking frame or other support to minimize the danger.

A slow, shuffling gait is a frequent characteristic of PD. It is due to the patient's asymmetrical shifting of their weight from one side to the other. There can also be difficulty or hesitation in initiating walking, as well as an increased risk of falling in those patients with postural instability. Although the optimum medication is of prime importance, physiotherapy can help reduce the motor handicap, especially by teaching the patient techniques that help control weight shift and also to overcome his difficulty in initiating walking.

Non-Pharmacological Treatment

Non-pharmacological treatment can contribute to reducing the handicap and improving the quality of life of sufferers and their helpers. They obtain the best results when they form an integral part of a program of treatment in the context of a multidisciplinary team. Physiotherapists can restore functional capacity to a maximum and minimize secondary complications thanks to reeducation of motor skills. Occupational therapists can ensure a maximum benefit for the patient from his personal work or leisure activities for as long as possible. Speech therapists can help counter speech defects resulting from loss of motor control. Nurses are needed to deal with incontinence in advanced cases. Dieticians help minimize weight loss that is often a feature of PD. Psychologists and psychiatrists can assess and treat mental symptoms either related to the disease or the medication. It is easy to see that such teamwork is necessary to maintain a minimum quality of life, but also what a burden it implies for the patient and his family, both materially and psychologically. The patient may seem to no longer exist except as a function of his disease.

In view of the obvious limitations of pharmacological treatment of PD, and stimulated by progress in surgical technique and our knowledge of the anatomy of the brain, there has been a rapid development in surgical treatment when symptoms become a real handicap. This can consists of the sectioning of certain connections in the brain. It can be used to treat tremor, "on/off" effects, and dyskinesia, but is of relatively little benefit for problems of balance and gait. It can, however, provoke negative cognitive changes, and carries the danger of damage to vision, speech and swallowing.

Deep brain stimulation (DBS) has similar effects to those of other forms of surgery, with the advantage of being reversible if necessary, and its effects can be modulated with time. It consists of implanting electrodes in the brain. We now have good long-term data on reduction of tremor with DBS, and the benefit seems durable. Observation of patients treated by DBS over ten years has demonstrated major improvements allowing medication to be reduced by 50% in some cases. DBS provides a good example of the utilization of advanced imaging without which it would be inconceivable. When medication is ineffective, DBS can be used to electrically stimulate a precise region of the brain through tiny electrodes. The target is the subthalamic nucleus, a small group of neurons, so-called because it lies just below the thalamus. Stimulation with small electric impulses has resulted in the symptoms disappearing in 70% of cases. It is analogous to a pacemaker, as used in the heart: an electrode is placed in the required area and connected to a small battery that sends electrical signals at regular intervals, either inhibiting or exciting this nucleus that influences the tremor. So, it is just as much a palliative treatment as is medication, not curing the disease but only reducing its symptoms.

A problem faced by signal processing engineers is that the subthalamic nucleus is unfortunately not visible by traditional MRI. The electrodes must be in exactly the correct position or the procedure will be fruitless: a small error in localization is immediately obvious, for the tremor does not disappear.

Improving Surgical Precision

The subthalamic nucleus is only 3 or 4 mm long and is deep in the brain, and invisible to normal imaging techniques. To rectify this deficit there exist other interesting, and surprising, resources. For example there exist atlases of the "standard" human brain, rather like geographical atlases, except that these days the format tends to be computer files rather than a book. The best known atlas is that of French neurosurgeon Jean Talairach (1922–2007), published in 1988. This "standard" brain is in fact not an "average" brain but is derived from the left hemisphere of a 60-year-old woman who had donated her brain for dissection.

Such atlases furnish information on what we know about the brain in general and, in the present case, where we can find the subthalamic nucleus. This information is vital, even if there is variation from one person to another. The work of the last several years has aimed to exploit to the maximum this information. This involves image processing, associated with lining up images from the atlas with the real anatomical image of the patient by a technique of deformation, so that visible structures in both help define the location of invisible ones. This technique is now sufficiently reliable to be used in surgery.

What we have discussed so far is a preparatory procedure for localizing the target nucleus. It serves no useful purpose to determine to the nearest millimeter where the target is, if we cannot direct the electrode to exactly that position. One method to obtain the precise target is to introduce a guide for the electrode by relying on a metal frame fixed to the patient's skull, which is imaged at the same time as the brain and which serves as a landmark. The coordinates of the target are recorded in relation to this frame, and by using various removable attachments the electrode guide can be placed in the target nucleus. The accuracy of the placement is checked immediately, for the patient need not be under general anesthesia and can be asked to cooperate in verifying the effect of the operation.

Future Perspectives

Another technique, that has been attempted over a number of years is to replace the degenerated neurons by transplantation of fetal neurons or stem cells. Results so far have been inconsistent, but in a small number of cases positive results have encouraged the pioneers to refine and better control the process. As the provision of enough fetal cells at the same time to make this type of transplantation possible is extremely difficult, recent efforts have been concentrated on the use of cultured cell lines and neuronal precursor (stem) cells. It is known that the stem cells involved in perinatal development of the brain are still present in the adult brain, but that their growth is inhibited. The processes by which this inhibition can be removed, as well as methods to replace growth factors, form the object of active research.

Study of the genetic changes associated with cell death in the brain provides another opening for avoiding or halting the process of degeneration. Research efforts include *in vivo* methods using viral vectors or direct transfer of DNA.

Some years ago the first step would have been to reproduce the disease in an animal. One method was to use neurotoxins, chemicals that are injected directly into the structure concerned to destroy its cells. But one was never sure that this really reproduced the process active in the human disease.

Today we dispose of genetic mapping. The first gene related to familial PD (about 5% of all PD) was recognized in 1996. Since then scientists have cloned more genes related to the disease. Cloning means that they have been identified such that we recognize the gene and its mutation. When a gene mutates there are two possibilities: either the mutation makes the protein that is manufactured by that gene toxic, or else it loses its function. Depending on which is the case different therapies are possible. If the protein is toxic we can intervene at the level of the messenger RNA, that is the intermediate step between the gene and the protein molecule that it manufactures, so that the protein is no longer produced. If on the other hand its function is lost, the protein must be replaced.

The priority is to understand the genetic basis, to make an animal model, and then develop a form of therapy. If this development is based on an animal model of which the genetics are understood there is a higher probability that the therapy will be effective in man.

Clinicians endeavor to identify families afflicted by a genetic form of a disease, and then the genes involved. Once the genes are identified, molecular geneticists enter into action. They first study the gene in transgenic animals, that is to say animals in which the gene has been deleted or modified. It may be possible to begin with rodents and then progress to primate models before transferring the therapy to human patients.

At the heart of this research is a protein called alpha-synuclein. It was discovered that patients with sporadic PD, that is those who do not have a genetic basis, have deposits of this protein in their substantia nigra when their brain is autopsied. The logical step was to ask the question whether PD with a genetic basis might be due to a genetic abnormality that caused this protein to be produced. Indeed three families have been identified worldwide that suffer from such a mutation and also from PD. One might, then, postulate that this protein is toxic for neurons that produce dopamine. Attempts are being made to counter this toxicity of alpha-synuclein by modifying the gene so that it produces another protein. To do this a carrier virus is used, in this case HIV. That may seem strange, but HIV is capable of modifying the genetic make-up of a neuron because it acts on cells that do not divide.

This so-called *retrovirus* technique has been developed by Didier Trono of the EPFL in the treatment of blood disease. Patrick Aebischer's laboratory then pioneered its use in the nervous system of rodents. They removed the pathogenic genes from the HIV, but retained its ability to affect nerve cells. They then took the virus that contained the gene of interest and injected it directly in the substantia nigra where, in infecting the neurons, it delivered the gene. The novelty of their technique is in designing a PD therapy that targets not just the symptoms but directly the cause. To do that, it must prevent the appearance of the first symptoms in patients with a family history when they are 20 to 30 years old.

Another interesting possibility is the use of imagery. This is one of the rare domains in which we can study the metabolism of dopamine with PET. A marker exists that enables us to follow the dopamine pathway so as to reveal a reduction of dopamine uptake by its target cells before symptoms appear. Obviously, we cannot expect everyone to undergo a PET scan, unless there is a suspected genetic factor. At the present time we do not have biomarkers available. For example we would like to know if sporadic Parkinsonians have different levels of alpha-synuclein in their blood or CSF. There is some evidence that such is the case. If so, we could undertake systematic screening in the hope of a timely treatment.

We know that the same may be true for Huntington disease and some cases of Alzheimer disease, in which a gene has been identified. Indeed, various laboratories are now working on ways to control the Huntington gene using a similar method of viral vectors. For all these diseases there is now an amazing prospect: sufferers condemned to a slow decline without a perspective of curative treatment can begin to see a sign of hope.

The problem that subsists is to go further. As an example, take amyotrophic lateral sclerosis (ALS), a form of motor neuron disease, also called Charcot's disease. It is characterized by death of the motoneurons in the brain and spinal cord, which control the body's muscles. The patients lose control of their muscles little by little, generally beginning by the limbs and gradually spreading. When the vital muscles, such as those of respiration, are affected, the patients die, fully conscious of what is happening. It is one of the most distressing diseases imaginable.

It is difficult to intervene in the actual neurons involved. The target nucleus may be so large as to need too many injections, like the striatum, the affected structure in Huntingdon disease. In contrast, the substantia nigra of PD is small, a few millimeters across, and therefore fairly easy to target with a small injection. The brain does not feel pain, or even touch. Local anesthesia is only necessary for incising the skin and penetrating the skull, and so the operation can be performed on the awake patient. With a tiny needle it is possible to inject 10 microliters of a solution containing the virus, enough to infect a million neurons. These experiments have progressed far enough now to envisage using the method clinically on humans within a few years. In contrast it is inconceivable to use this technique in ALS.

Because in some cases PD has a genetic basis, it would be possible to systematically use medically assisted procreation with a preimplantation diagnosis, so as to eliminate the gene once and for all. This would be conceivable in families with hereditary PD, but the number of genetic neurodegenerative diseases is put at more than 5000. Unless some fundamental breakthrough is made in basic research, medicine will never be able to cure them all at the individual level.

The story of HM, the man with the split memory

"HM" suffered from severe epilepsy since infancy and in 1953, when 27 years old, underwent an operation by surgeon William Scoville in Hartford, Connecticut, to remove part of his temporal lobes, including the hippocampus on both sides. Neurosurgery was often the only method available to treat severe intractable epilepsy. At the time, knowledge of cortical function was less advanced than now, and his surgeons could not have foreseen what would happen.

After the operation HM's epilepsy was much improved, but he was unable to recall events older than a few minutes. For example, he could carry on a completely normal conversation with a stranger, but not recognize him five minutes later when the person had been away to answer the telephone.

(continued)

(continued)

HM kept fairly precise memories of much of, but not all, his past up to about three years before the operation, but he was incapable of remembering anything since then. We call this condition anterograde amnesia, meaning loss of memories since the lesion. HM could use the telephone, so he could remember things for a few seconds, or even minutes. He could ride a bicycle, and so was able to use learned motor behavior. He could also speak, write and read properly, suggesting that his access to memory for words and grammatical structures was intact, although minutes later he would read the same text again thinking it was the first time he had seen it. Although HM's case was catastrophic for the patient, except that it helped the epilepsy, it is very instructive in showing that our memory is divisible into several types, and something about the localization of one of these types.

Chapter 6
The Treatment of Alzheimer Disease

Alzheimer disease (AD) is a degenerative disease causing lesions in the brain, and is the principal cause of dementia. It is named after German neurologist Aloïs Alzheimer (1864–1915) who first described it in 1906. Certain brain cells shrink or disappear to be replaced by irregular dense patches called plaques, containing a protein, amyloid beta (Aβ). The Aβ molecule is a fragment derived from a protein called amyloid precursor protein (APP) which is embedded in the wall of a cell, and of which the outer fragment can be cut by enzymes. This is a normal cellular activity throughout the body, but it can degenerate. Aβ is not itself toxic, but is when it forms molecular aggregates. If the concentration of Aβ is too high, the molecules agglutinate to form the plaques that then stifle the neurons.

Another diagnostic feature of AD is the presence of neurofibrillary tangles, aggregates of tau protein, again triggered by Aβ, but this time in the cells, which modifies enzymes that add phosphate to tau proteins. The process finally destroys healthy neurons.

Although the disease causes loss of many functions it does not change the patient's capacity to experience feelings of joy, anger, fear, sadness or love, and to react accordingly, at least in the early stages. It is impossible to restore normal function to the damaged brain cells, but nevertheless there exist forms of treatment and strategies that can help the patient and his entourage. Unfortunately the diagnosis is often made very late, for the patient does not realize he is ill, unlike in PD when the sufferer quickly notices his motor difficulties.

The commonest form of AD is sporadic, and accounts for 90 to 95% of cases. Its causes are unknown, but the greatest risk factor is age. In a rarer familial form, 5 to 10% of cases, there is genetic transmission from one generation to another. Estimates put the worldwide incidence of AD at 25 million, and this is set to triple by 2040. Some 10 to 15% of cases occur between 45 and 65 years of age. However after age 65 up to 5% of the world's population are affected, and 20% after age 85. In developed countries AD is the fourth cause of death after cardiovascular disease, cancer and stroke. As the population of these countries is constantly increasing in age due to the progress of medicine in the prevention and treatment of other diseases, AD confronts us with a formidable challenge for the century to come, simply in terms of the costs involved for palliative care and other treatment.

J. Neirynck, *Your Brain and Your Self: What You Need to Know*,
© Springer-Verlag Berlin Heidelberg 2009

The Consequences of AD

The limbic system is under attack from the very beginning of AD. It is normally involved in emotion and helps integrate memory and behavior. It also controls certain day-to-day functions such as feeding and sleep.

In the end, AD affects all aspects of a person's life: thoughts, emotions and behavior. Every patient reacts differently, and it is difficult to predict the symptoms that a given individual will manifest, as well as the order in which they will occur and the speed with which they will evolve. AD interferes with mental capacity, such as comprehension, communication and decision making. Patients find it impossible to accomplish simple tasks that were familiar for years, cannot find the right word or follow a conversation, and lose interest in leisure activities. Memory loss is at first for recent events, but later for the more distant past. With it go symptoms such as the repetition of gestures and words, but also agitation, physical violence, and then a gradual physical decline. This deterioration has a profound effect on a person's ability to accomplish even the simplest day-to-day activities, such as eating, without help.

Today's Treatments

It is impossible to cure AD, but we do have medication to help control the behavioral symptoms and to treat the cognitive deterioration.

Cognitive deterioration can be helped by drugs that increase the levels of the neurotransmitter acetylcholine (ACh), thus improving communication between important sets of neurons. In addition to simply increasing the concentration of ACh in the brain, there is hope now of being able to manipulate certain receptors that fix the incoming transmitter molecules at a synapse. Excessive stimulation of so-called NMDA receptors can cause neuronal death. Recent studies have shown that memantine, which blocks this excess stimulation, can slow the progression of moderate to severe AD. Further, estrogens, anti-oxidants and anti-inflammatory drugs may slow the appearance of AD. Possible beneficial effects of vitamin E and deprenyl are also being studied. Research suggests that good food hygiene also contributes to slowing the progress of the disease.

Dementia is a complex condition with multiple symptoms in addition to deficits in memory or functional capacity, including depression, anxiety and sleep disturbance. There may be psychotic symptoms such as visual and auditory hallucinations (seeing non-existent objects or hearing voices) and even imaginary smells and tastes. There can be delusions, such as being convinced that things have been stolen, and mistaken identities, such as believing that someone close has been replaced by an imposter.

Many people suffering from dementia also manifest non-cognitive symptoms at some stage, often referred to as Behavioral and Psychological Symptoms of Dementia (BPSD). Their treatment is at least as important as that of the cognitive ones because they can be a source of intense stress and anxiety both for the demented patients and for their carers. BPSD form an essential factor in making decisions related to long-term treatment such as intensive or institutional care.

The first stage is to evaluate the nature and temporal pattern of the symptoms. BPSD can be worsened, or even caused, by a parallel physical or environmental condition. Thus, a urinary infection can lead to sleep disturbance, influenza can accentuate depression, arthritic pain can cause anxiety and aggressivity, and a chest infection can precipitate psychotic symptoms. In each case the appropriate treatment is to tackle the underlying pathology and not the BPSD. Among environmental factors triggering BPSD is a change in daily routine, the arrival of a new carer, or even a new treatment. In such cases it is important to modify the environment to make it more acceptable, to help a carer adapt to the patient, or to create an atmosphere of entertainment or occupational activities.

In many cases treatment of a physical condition, modification of environmental factors, or behavioral therapy, suffice to relieve BPSD. Nevertheless drug treatment is still often necessary, in which case the regime must be consistent with a minimum disturbance for the patient and must be based on medication that is designed to be well tolerated by elderly demented patients. For example, many psychoactive drugs, including tricyclic antidepressants, have an anticholinergic action and might adversely affect cognitive ability of AD patients in whom cholinergic activity is already depressed. Among the drugs used in treatment of BPSD are the antipsychotics, but they must be used to treat specific symptoms and not simply to tranquilize demented patients. Their use must be reviewed regularly and the best practice is to use them only when necessary and at the lowest possible dose and for as short a time as possible. Other drugs are in use to control BPSD and sleep disturbance, but many physicians judge that such medication can increase confusion in the demented, and avoid their use.

Research for New Therapy

Today's treatment of AD is merely palliative, but research is in progress to tackle its cause. There are numerous clinical trials being undertaken throughout the world to validate treatment that will only be commercially available after 5 to 10 years.

Although familial AD forms only a small fraction of cases, it has provided a key to a possible curative therapy. Various genes have been identified as associated with familial AD. One codes for synthesis of the $A\beta$ that aggregates to form the characteristic plaques that cause neuronal death. Even in the sporadic form the same mechanism is involved except that there is no genetic basis for the production of $A\beta$. Environmental factors could also play a role, especially vascular factors such as hypertension, excess cholesterol and tobacco, as well as stress. On the contrary, physical and mental activity, hormonal balance and an appropriate diet can reduce the incidence of AD.

The most obvious line of research would seem to be to avoid the formation of $A\beta$ by inhibiting the enzymes that detach it from APP. It has been discovered that it is possible to introduce in the brain relatively small inhibitors that easily cross the blood-brain barrier. As an example, the Eli Lilly company is testing a drug of this type that has already passed Phase I of clinical trials on healthy volunteers to ensure its lack of toxicity. It is now entering Phase II which consists of treating patients with AD. Another drug, fluizan, shortens Aβ which inhibits its agglutination. It has reached phase III of clinical trials on around 1000 patients.

Another line of research concerns attempts to clear the brain of plaques once they have formed. The firm AC Immune in Lausanne is confronting the toxic protein of plaques. This protein has an aberrant form, not recognized by the immune system. The approach consists in introducing into the patient's body fragments of proteins that stimulate the immune system to produce antibodies and thus dissolve the plaques. By introducing only fragments of proteins, the inflammatory effects of whole proteins are avoided. After *in vitro* experiments demonstrated that the antigens effectively reconfigure the pathogenic proteins, tests on mice have confirmed that plaques dissolve in a third to half the animals.

Pierre Magistretti at the EPFL is pursuing yet another line of fundamental research. His main field is cerebral metabolism and the mechanisms that ensure the interaction between neurons and glial cells. For long it was believed that the metabolism of the brain was similar to that of muscle. But this view does not take into consideration the functional specificity of the brain. When you stimulate the visual system the parts of the brain that process a visual signal consume more energy. Where does it come from and how does it reach the neurons? In fact this is one of the functions of glial cells, that have attracted much less attention than neurons. They were often considered as simply ensuring the mechanical support of neurons. Glia contain receptors for neurotransmitters and they participate in the intercellular dialog of the brain. We recognize macroglia and microglia. Of the macroglia, the oligodendrocytes make myelin to insulate axons. In multiple sclerosis this myelin sheath degenerates because the oligodendrocytes die.

Another type of macroglia is the astrocyte, cells that harmonize the energy of synaptic activation. They contact the blood capillaries of the brain and pass glucose to neurons. A recent discovery links astrocytes with synaptic activity and glucose uptake. Neurotransmitters, such as glutamate, act only for a matter of milliseconds, after which they are taken up, partly at least by astrocytes. Glutamate uptake triggers a need for energy. Lactate leaves the astrocyte, and is converted to pyruvate that can then produce energy. So the astrocyte enables the coupling of synaptic activity with glucose uptake and its supply to the neuron. In summary, brain activity depends on neuronal function, which in turn depends on triggering an energy pump represented by glia.

If neurons do not receive an adequate energy supply because the coupling mechanism is no longer active, they will be much more vulnerable to all sorts of aggressions. This can contribute to, or participate in, the mechanism of neuronal death. This energy hypothesis represents one of several explanations of neurodegenerative disease in general, such as AD, PD and ALS. Faced with problems of energy supply, neurons would suddenly become vulnerable to other factors. So in persons genetically susceptible to AD we might detect decreased metabolism well before the cognitive symptoms.

This being so, one approach to AD would be to develop drugs to improve glial energy metabolism. We now have *in vitro* results indicating that we can make neurons more resistant to chemical aggressions by upregulating the levels of transporters of glucose and lactate in astrocytes and neurons respectively. Improving energy metabolism results in a neuroprotective effect.

The story of Ravel's secret opera

The first works of Maurice Ravel (1875–1937) were performed for the Parisian public in 1902. He rapidly became one of the most performed French composers, closely following Claude Debussy.

In 1927 he began to complain of neurological troubles: fatigue, amnesia, neurasthenia. In 1928 he returned from a triumphal tour of America and composed his *Bolero*. In January 1932 he directed his own piano concerto in G major. In October of the same year, he was knocked over by a taxi without, apparently, serious injury. But, from that moment, he sank into perpetual somnolence. In June 1933, he found himself unable to make swimming movements. A year later, he needed a dictionary and an enormous effort to write a simple letter.

From then on Ravel lost his faculties one after the other. He could not coordinate his movements, fix his gaze or speak clearly. He was incapable of composing music, although he felt the inspiration. He never completed either of his projects for the ballet *Morgiane* or the opera *Joan of Arc*. In his own words the opera was in his head but he could no longer put it into music. His neurological disease had extinguished his creativity.

Ravel underwent neurosurgery on December 17, 1937, a risky procedure in view of the relatively primitive conditions at that time. He died a mere ten days after this ill-advised operation. There was no autopsy.

We do not have a precise diagnosis of Ravel's disease. We can, however, exclude a stroke or a sequel to his traffic accident. We equally exclude AD, for Ravel's memory remained intact, as did his lucidity. Today, some specialists classify his condition as symptomatic of Pick's disease, a dementia with different characteristics from AD, associated with degeneration of the frontal and temporal lobes.

The most fascinating thing in this story about Ravel concerns the loss of those masterpieces that he would have created had his brain not betrayed him. His inspiration endured. So where was this inspiration located?

Chapter 7
The Cerebrovascular Accident

The cerebrovascular accident (CVA) can be defined as the sudden appearance of disturbed cerebral function of which the cause is a vascular event. This can be the rupture of an aneurysm, a swelling on an artery as the result of a congenital malformation, that bursts suddenly and unexpectedly. It can also be the sudden culmination of a slow decline in brain blood flow due to lowered blood pressure, but equally possible is the bursting of a cerebral artery due to high blood pressure. Then again, the cause can be thrombosis, the blockage of a blood vessel by a clot or a cholesterol plaque within the brain. An embolus is a similar clot produced elsewhere in the body then released into the circulation and ending by blocking an artery in the brain by bad luck.

Deprived of oxygen, the brain degenerates in the region supplied by the artery in question and neurons begin to die. Functions such as speech, movement, vision and memory can be affected in minutes. If the symptoms disappear within twenty-four hours we call this a transient ischemic attack (TIA). The CVA, or stroke, is the main cause of acquired neurological handicaps in the adult, throughout the world. It is not just a disease of developed countries: the same problem affects developing countries, although the precise forms may differ.

A Typical Stroke

The typical CVA comes out of the blue. Certainly, the patient may well present a number of risk factors, but he is probably in top form, and this is a sudden and acute blow. It diminishes his sensory and motor functions, and also his higher faculties of perception or the capacity to express himself. A dramatic problem brought on by a stroke is a decreased ability for judgment. Whereas it is an affliction that demands immediate and urgent action so as to minimize death of brain cells, the patient may not realize what is happening to him. In the case of that other major vascular accident, myocardial infarction, or heart attack, the patient suffers violent pain and senses that he is about to die. He will immediately seek medical aid. On the contrary, the altered state of awareness related to a stroke may manifest itself as trivialization or denial of the problem. It can even be contagious: the family may well postpone

J. Neirynck, *Your Brain and Your Self: What You Need to Know*,
© Springer-Verlag Berlin Heidelberg 2009

action and even the physician may leave the patient with the advice that all will be better tomorrow. This is clearly the error not to make, for tomorrow nothing will be possible to counter the effects of a CVA.

Patients do not realize what is happening. If, for example, they find difficulty in moving an arm they do not relate it to an incident in the brain, but prefer to see it as a trivial problem of the circulation in the arm. If someone is paralyzed in half their body, one would expect them to be concerned, but often it is not the case.

Classic Therapy

So, the first thing is to recognize the CVA. The patient is examined and may be subjected to a number of tests, such as CT, MRI or angiography, in order to determine if part of the brain is devascularized. Once that is determined, it has to be decided whether the affected region is already dead or whether there is a chance that brain cells can still be saved. If the latter is the case a number of treatments can be instituted, including the dissolving of a clot blocking a blood vessel by means of medication. Time is the key element, for the window of opportunity for effective action is closed after three or four hours.

It is also possible to envisage more invasive techniques, that may not involve neurosurgeons, but rather neuroradiologists. They introduce a tiny catheter into a blocked artery until it reaches the clot, at which point they attempt a mechanical ablation of the obstacle.

CVA and Higher Functions

After a stroke, a patient may develop aphasia, that is to say loss of speech, either because of an inability to express himself or to understand language or, often, both. This loss of communication, either in terms of expression or comprehension, also makes difficult the understanding of what had happened, as also the necessary explanations upon arrival in hospital. On the contrary, it is rare that a CVA can be mistaken for an acute psychiatric condition for characteristically a stroke does not affect all cerebral functions, but a restricted part of the brain that may not be involved in higher mental activity. However some time after the acute stroke, patients may develop severe depression, especially when they realize the seriousness of their condition. Nevertheless it could also be that parts of the brain are involved that do play a role in the development of depression.

What forms of rehabilitation exist if part of the cortex is damaged? They are very limited and are based on good intentions more than scientific reality. There is much talk about very little, for we just do not have effective means. In the present-day state of medical knowledge we cannot hope for new neuronal growth, nor can we reorganize surviving cerebral tissue in new ways that would enable them to take over lost functions. Patients must be maintained in the best possible condition by the use of physical means such as physiotherapy and logopedic training, that is the repetition of the same words in the hope of recovering an embryonic form of language. However, full recovery is seldom possible in spite of ever improving effectiveness of these methods.

Current practice is aimed, rather, at prevention at the cerebral level, or at the level of general health. If, in spite of preventive measures, a stroke does happen, it is vital to unleash the available therapeutic arsenal as soon as possible in order to have a hope of any functional recovery. If, after all, there are clinical sequelae there is little chance of completely regaining lost function. Medication may be used to attenuate secondary effects such as pain or muscle spasm, but will not provide a cure in the real sense of the word.

Prevention

Prevention must be envisaged for patients at risk, such as those with evidence of arteriosclerosis, a thickening of arterial walls by cholesterol or other substances. There are various measures that can be taken such as:

- *Primary prevention* in patients at risk, but who have not yet shown signs of complications, such as those with hypertension or high cholesterol, or who have evidence of arterial malformation. Depending on the details of the case, one can either prescribe a drug to prevent further deterioration, or undertake surgery to, for example, widen an affected artery or remove part of it.
- *Secondary prevention* in patients who have already suffered a clinical episode but who have recovered relatively well and in whom it is desired to prevent a relapse. Such prevention can be successful even if not undertaken very early in the process.

It does not seem unreasonable to envisage a regular check-up for the whole population after about 50 years of age. It would not cause galloping inflation in the cost of health care if limited to a few key investigations, say every two years initially and every year at a later age. In contrast, systematic prescription of medication to control excessive cholesterol is expensive, is not effective in all cases, and can cause undesirable side effects in some patients.

Hope from Research

Stroke research is mainly aimed at the acute phase. Radiologists are working to image the brain during the events taking place soon after the vascular accident. Images three or four hours after a stroke look fairly normal and are of little clinical value except to exclude other pathology. Now methods are being developed that not only visualize the lesioned brain tissue, which may be undergoing necrosis, but also distinguish areas that are still recoverable if, for example, a blocked artery can be unblocked. In some cases there is already irreversible damage after a few minutes, but it has been seen, using this technique, that even after three hours or more there are quite large areas that can have either of two outcomes: they may evolve in the direction of necrosis but, if the artery responsible for supplying the area is unblocked immediately, they may recover.

This imagery is based on a perfusion CT scan. A contrast medium is injected and its passage followed through the various components of the brain. Then the captured images are analyzed by computer to distinguish, by color coding, permanently damaged areas from areas that could evolve in either way. If there is a lot of green, a fairly aggressive treatment might be indicated, but if red predominates treatment will be relatively contraindicated. One could use MRI in the same way but CT is much easier to employ and more rapidly available at short notice, day and night. It is not a complicated technique and fairly easy to use routinely. So it can be used for a patient arriving in the emergency room at 3 am without the intervention of a large expert team.

In the long term the aim is to protect neurons which have a reduced blood supply. This ischemia leads to a series of metabolic events such as penetration of certain ions into the cell. This results in electrical activity which stimulates the production of various molecules and inhibits others. This metabolic cascade ends with the death of the cell. The idea is to inhibit the principle mechanisms for at least sufficient time to unblock the artery.

At present the best protection might be hypothermia, like putting the patient into hibernation, but unfortunately it is usually technically impracticable as it would require an enormous infrastructure. It is not enough to simply lower body temperature; one must envisage intensive care and the monitoring of vital functions which could be endangered by the treatment. In other words, this is a treatment that can be considered for special cases but cannot yet be applied widely for economic reasons.

Another long term solution is rehabilitation. What can we do to modify the function of surviving cerebral tissue? Can we encourage regrowth of neurons where they have been lost? Theoretically we could, but in technical terms we certainly do not have this option yet. The risks are great and there is no current clinical application available. It is not at all the same thing to study stroke provoked in a young laboratory rat and to treat an elderly patient who has a multitude of risk factors and other problems.

Statistics show that such approaches could be effective in terms of a reduction in mortality, that is somewhere around a third to a half of cases depending on age. The effect of prevention is positive and spectacular. In some countries where hypertension was poorly treated, once programs were set in motion to rectify that, the mortality due to stroke was halved. However, as aging of the population is unavoidable the absolute number of cases tends to increase rather than decrease.

Stroke will remain a major cause of morbidity and mortality for the foreseeable future, in contrast to certain other brain diseases. It is impossible to prevent it definitively. The situation will certainly slowly improve but there will be no sudden radical and decisive changes, as antibiotics decisively changed the prognosis in infectious disease such as tuberculosis or syphilis.

Rehabilitation

Neurorehabilitation aims to recover as much function as possible after lesions that have caused even major damage to the CNS.

When the brain is damaged, either by a stroke or by mechanical trauma, the first thing to determine is which cognitive domains, if any, show deficits. For example, when the lesion is in the left hemisphere there is usually a language deficit: the patient has difficulty in speaking or understanding, or has trouble in a particular aspect of language. It is important to decide what is not functioning, as that will determine the potential for recovery. There are certain language deficits that almost never recover, and which necessitate strategies of substitution. In other cases there is a reasonable hope of recovery of some degree of language function. To establish a diagnosis one can evaluate language itself, but also other cognitive functions such as memory, visual recognition, and the ability to plan and organize. In some patients one can study preserved cerebral function in various domains by functional imaging in order to determine which of those domains might serve to rebuild function.

Later, different types of treatment exist. Logopedia involves intensive reeducation in an attempt to revitalize whatever neuronal circuits might still reach a functional level. Sometimes it is possible to use related functions that have been spared, such as singing, to find an entry point for reeducation of speech. In other cases computer technology can be helpful, or the patient can be exposed to an aggressive therapy in which he is forced to use language. For example, a patient who cannot say that he wants to eat may be able to sing, even complex melodies, although he cannot formulate a spoken sentence. The therapist can use a melodic prop to express, "I want to eat". Patients sometimes manage to sing this, and may even finish by being able to say it. It is a technique that has proved itself but is not widely used.

When patients have a left hemiplegia after a cerebral lesion they may be able to learn to walk again. It is extremely important to make a diagnosis and determine where exactly the problem lies in order to formulate a rehabilitation program. Physiotherapists can then elaborate a carefully targeted approach for an individual patient. A patient who succeeds in repeating a movement thousands of times is likely to recover better than one who does it only a few dozen times. From the moment that such reeducation is instituted, certain brain areas are slowly reinforced by repetition: one could speculate that the cortical area used in singing will slowly be taken over and used for speech, but this is still far from clear and is a subject that needs research to find an explanation.

In most right-handers language is represented in the left hemisphere. The right hemisphere contributes, but the left is the main player. When there is a lesion in the language areas of the left hemisphere, the right can help. This help may well be temporary, and even if it is permanent it often does not represent a high degree of recovery. On the other hand, studies on the use of singing have shown that this method can facilitate recovery after left-sided lesions affecting language.

An important feature that may cause problems for patients with cerebral damage is heminegligence. This is an astonishing syndrome in which the patient pays no attention to what happens on the side opposite his lesion. If the lesion is on the right, he may fail to eat what is on the left side on his plate, or not read the left side of his newspaper, even if he realizes that the food is there and that the page contains news. He pays no attention and does not even turn his head to the neglected side. This is a really crippling syndrome in everyday life. Sufferers respond poorly to physiotherapy. The prognosis is bad.

Heminegligence can take various forms. The traditional specialized reeducation consists of attracting the patient's attention to the affected side. There is no suggestion of a loss of vision on one side. A patient who is blind in the visual field of one side after an occipital cortical lesion will not suffer from heminegligence for he will compensate fairly easily. The lesion is in fact in the parietal or prefrontal cortex, regions that relay from other cortical areas and contribute to a network that seems to be involved in attention and initiation of activity.

Reeducation of memory is a complex affair, for we possess several types of memory, as we saw in Chapter IV. First there is short term memory, like remembering a telephone number read in a directory long enough to dial the number. It operates for a few seconds. Then there is the memory we use to describe or draw an object, a person or a scene that we saw some hours, months or years before. This is the memory that is most often affected by cerebral lesions, causing what we term amnesia. Finally, we use procedural memory, such as when we learn to ride a bicycle, or to ski, or to read mirror image writing.

These different memories can be more or less selectively affected by cerebral lesions, but it is mainly long term memory that is affected most often, and which is the most disruptive for everyday life. One can try to reeducate the patient to use various strategies. Long term memory depends on specific brain regions, such as the hippocampus and related structures. When they are damaged there is usually no hope of retraining the patient's capacity to store new memories, but one can offer substitutes, such as computer assistance.

The ability to recall a memory or to recognize something that I see is the result of activation of a neuronal network that can be spread widely throughout my brain. Is it due to a mere activation, or to a discharge frequency, or to coincidental discharges throughout the network? Several hypotheses support a temporal pattern of discharges within the network as being of vital importance. So, learning a poem means that it is remembered when a network of synapses stores that particular memory. But we must distinguish between the network that enables a memory to be formed and the place where this information is finally stored. The hippocampus is the organ that enables long-term memory to be stored, but the information is not stored in the hippocampus. Patients whose hippocampus has been damaged can still remember things up to a few hours before the lesion.

It is the same with anesthesia for a surgical operation. The patient will have partial amnesia for the few minutes before the start of the anesthesia because the drugs used prevent transfer from the hippocampus to the long term memory store. In the same way, the last few minutes before falling asleep naturally are not remembered: you never remember actually falling asleep. The way in which this transfer from the hippocampus to the permanent long term memory store occurs is the subject of current research. It is limited to animal models, for it is impossible in man in whom we have no access to the mechanism.

The story of Valéry Larbaud, the repetitive poet

The French poet Valéry Larbaud (1881–1957) was born in auspicious circumstances. He inherited a fortune that enabled him to write freely without having to worry about earning a living. He founded a literary style, was a cosmopolitan traveler, translated James Joyce, revealed foreign literature to the French public, and was the friend of numerous authors of his generation.

In August 1935, Larbaud was crippled by a stroke that left him hemiplegic on his right side and totally aphasic. He survived 22 years after his stroke, during which he was never again able to write or dictate anything.

His aphasia evolved from complete mutism to a characteristic pronunciation of a single phrase, not without its charm and significance, "*Bonsoir, les choses d'ici bas*" ("Good evening, things down here"). Thus he replayed the stereotyped syndrome that is fairly common in the circumstances and that had affected another great French poet, Charles Baudelaire, in 1866 with the difference that the latter, being much less well brought up, only managed to exclaim "*Crénom!*" (roughly, "For God's sake!").

This single phrase that Larbaud could pronounce worked as a visiting card for a while. It was perfectly articulated and spoken quickly as if he were not concentrating. Sometimes Larbaud embellished it with a little laugh, as if conscious of the incongruity of what he was saying.

Later he managed to express his emotions. Recalling his youth, he repeated: "Ever marvelous youth". He relapsed into a telegraphic style. For example he might illustrate an encounter by "Today good evening talk literature". He managed to organize his works for editing in Parisian editor Gaston Gallimard's famous *Pléiade* series of books by telling him that a given book was "No good". We can thus assume that his memory and judgment were intact, even if he could not communicate them in words. This is a classic case of aphasia.

Toward the end he sealed himself off in total mutism, refusing to receive visitors, reading only a dictionary. The writer had vanished into his own brain's defect. On his death, he whispered "Thank you" to the woman who had been his companion in life.

Chapter 8
The Fatality of Tumors

Like all tumors, those in the brain result from uncontrolled proliferation of cells. There exist primary tumors that form directly in the brain, and secondary tumors, metastases of primary tumors elsewhere. There is a great variety of primary tumors of very varying malignancy according to the type of cell and the site. The frequency of these tumors, for example in the USA, is around 6 cases per year in a population of 100,000, with men being more often affected than women. One in six cases is in patients aged less than 30, a third are at ages 30 to 60, and half at more than 60. The overall mortality from these tumors is around three quarters of all cases, but it varies widely according to the type of tumor. Only one third of patients survive for five years after the diagnosis. Brain tumors are the first cause of death in young people under 20.

We know little about the origin of primary brain tumors and so cannot give much advice about their prevention. Some seem to involve a genetic factor. We know that children who have been exposed to radiation have a high incidence. Concerning secondary tumors, cancers of the breast and lungs often send metastases to the brain.

As brain tumors, like others, grow by multiplication of cells, the intracranial pressure increases and compresses healthy brain cells which may die. Vital functions, such as vision and motor control, may be threatened by this effect. The raised pressure may cause headache or epilepsy, as well as troubles of memory and intellectual ability. If a brain tumor is suspected on account of particular signs and symptoms, the diagnosis is established by cerebral imaging which will localize the tumor, and perhaps also by a biopsy performed through a fine needle in the growth.

Treatment of a primary tumor will usually be by surgery, often supplemented by radiotherapy or chemotherapy if they are malignant. Progress in surgery often allows the complete removal of a tumor with minimal damage to surrounding tissues. But there are sometimes inevitable sequelae, such as the loss of certain faculties, perhaps a reduction of the visual field. Radiotherapy can improve the life expectancy of the patient, but there are well-known secondary effects such as fatigue and hair loss. Chemotherapy is difficult to administer because the blood-brain barrier prevents or hinders the absorption of the chemicals so they are sometimes injected directly into the brain.

J. Neirynck, *Your Brain and Your Self: What You Need to Know*,
© Springer-Verlag Berlin Heidelberg 2009

Classification of Tumors

A distinction can be made between meningiomas, growing from the meninges, the membranes that surround and protect the brain, which compress the brain but do not actually invade it, and other tumors that grow within the brain itself. Neurons are rarely the source of tumors, as they stop multiplying about halfway through fetal life. However, this is not an absolute rule. It has been discovered that deep in the brain, in the lining of the ventricles, there are adult stem cells that can regenerate neurons. But still they rarely give rise to tumors, although this can happen, especially in infants.

The commonest brain tumors are derived from a glial cell, the astrocyte. The World Health Organization has developed a grading system to describe the behavior of astrocytomas. Grade I astrocytomas, the so-called pilocytic astrocytomas, are most frequent in children and can be perfectly benign. They are the only astrocytomas that can be cured by surgery, for they are rarely invasive. In the adult, three other grades are described in ascending order of gravity. Grade II, the diffuse astrocytoma, differs from Grade I in being invasive, sending cells some distance from the growth itself. It is practically impossible to remove it completely by surgery and these Grade II astrocytomas regrow in 90% of cases. When they regrow it is usually with a higher level of malignancy, either III or IV. Thus these tumors evolve from low to high malignancy. There are, as was said above, many other types of primary brain tumor. Detailed information is beyond the scope of this book but references can be found in the Bibliography (page 115).

Molecular genetic studies have enabled the identification of certain genetic changes that are related to changes in malignancy. Research is in progress to block the molecular mechanisms that lead to this increasing malignancy. Genetic change over time in these tumors involves the loss of cancer-suppressing genes that slow cell multiplication, and the appearance of oncogenes. These accelerate the division of cells that finally runs out of control, letting the cells proliferate wildly and destroy the brain.

The Search for Treatment

This research is well advanced but so far has not resulted in the discovery of substances to block this genetic decline. Gene therapy consists of the modification of oncogenes by using viruses as vectors to introduce a gene that can block the genetic changes that cause cancer. This technique is thus similar to that described earlier for the prevention of Parkinson disease of genetic origin. This research still has no direct clinical application, but may have in 10 to 20 years time.

Another approach is to search for an explanation of why some tumors resist chemotherapy. The expression of a gene depends on a promoter. Those patients who do not respond to chemotherapy possess a gene for an enzyme that blocks the efficacy of the chemicals used. Other patients, about half in fact, have a promoter that stops production of this enzyme, and they respond well to chemotherapy. This enzyme is called MGMT (methylguanine-DNA methyltransferase). Patients who respond to chemotherapy have a methylated form of MGMT. They would hope to survive for two years or more, while the non-responders would die in a year or so.

Another important subject in cerebral tumor research is that of stem cells. In the embryo, stem cells divide and then differentiate into the various cell types that form the organs of the adult body. Stem cells have been found in cerebral tumors and the theory is that these stem cells allow the tumor to grow indefinitely. Although with the various treatments now available, such as radiotherapy and chemotherapy, we can destroy most cancer cells, we cannot destroy the stem cells, that survive to renew the cancer's growth, and the patient relapses.

A cascade of events in the stem cells leads to their proliferation. It may be possible to inhibit this process to stop the proliferation. The gene associated with this cascade has the picturesque name of sonic hedgehog, which reflects the innate humor of researchers. The cascade can be blocked at a particular stage by the use of cyclopamine. It is present in certain plants and has the peculiar property that animals that eat it tend to give birth to young with only one eye ("Cyclops"). The sonic hedgehog gene is therefore very important in the development of the embryo, and thus may be important in cancer. If one treats stem cells taken from a tumor with cyclopamine *in vitro*, the growth of the cells is inhibited.

Another line of research is the use of embryonic stem cells to treat cerebral tumors. Embryonic stem cells are capable of migrating over considerable distances and it seems that they are attracted by tumor cells. So a whole research domain is devoted to the use of stem cells as vectors to transport agents that might destroy tumors. Some researchers even claim that stem cells themselves destroy cancer cells directly, but that may be a pious hope rather than scientific reality.

Yet another research theme is immunotherapy, or vaccines against cancer. As we have noted several times already, the brain has a reputation, albeit a rather mythical one, of being protected by the blood-brain barrier which might prevent its immunological response being as effective as in other organs. In reality, the blood-brain barrier is not as water-tight as that, and it is possible to detect immune responses in the brain and to exploit them in the fight against cancer.

The story of two men without emotions

During the 1970s the first of these men, Elliot, was an excellent husband, a good father, and worked for a commercial company. He had attained a respectable social status, until he began to suffer from violent headaches. His doctor diagnosed a brain tumor, a meningioma situated just above the nasal cavity in the midline of the brain. As it developed, the growth compressed his two frontal lobes and precipitated a change in Elliot's character. He lost all sense of responsibility.

(continued)

(continued)

Elliot underwent surgery, involving the ablation of part of his brain. As the tumor was benign there was no regrowth. Although he recovered physically very quickly and his mental faculties remained intact, the same good news was not true for his character, for Elliot was incapable of managing his life. He shifted from one task to another in an unrealistic way. He undertook some jobs with an obsessional striving for perfection, even if it meant neglecting other urgent matters. In the end he lost his employment.

Since then he never managed to keep a stable job, and engaged in senseless financial speculation that ruined him. His wife left him and he was never able to recreate a relationship. Nevertheless, the social security refused to consider him a victim of any invalidity and refused any form of pension. Within his entourage Elliot was considered to have become lazy because his mental faculties seemed unchanged. He moved around normally, spoke clearly, was in touch with the news, and had an intact memory for all that had happened. Elliot was a victim of the dualist myth: because he enjoyed all his faculties, but still made erroneous decisions, he was responsible for his actions and deserved sanctions.

Elliot's salvation came from the neurologist Antonio Damasio in Iowa who was able to examine him and who attributed his change in character to neurological sequelae of the surgery. Elliot was perfectly capable of reasoning and predicting the disastrous results of his actions, but he could not draw any lessons from his repeated errors. He was like a relapsing delinquent who returns to his criminal behavior immediately after release from prison. He no longer exercised his free will; his ability to choose had become non-existent.

After undergoing a battery of psycho-technical tests, Elliot proved to be of above average intelligence. His short and long term memory, his capacity for learning, speaking and calculating, and his attention span were all intact. These are just the elements that are disturbed if the frontal lobes are significantly damaged. The trauma of his operation was limited: it affected his behavior but not his ability to evaluate events.

Nevertheless, during his sessions with Elliot, Damasio discovered something curious. Elliot did not suffer from his tragedy: he showed no sadness or impatience or anger. He had become incapable of emotion. He still remained able to evaluate a social situation and conceive of the consequences of his decisions, but he did not feel personally involved in them. His ability to choose and that to feel emotion had declined in parallel. He had become asocial, without having the remotest intention to do so. In the end, Elliot's story is of a missed opportunity: that of a definitive surgical cure of a benign tumor without collateral damage to the brain. Techniques of diagnosis and surgery have improved over the last 30 years.

To reassure today's patients, one can relate the case of Etienne, a young surgical trainee in Geneva. He presented character changes, similar to Elliot's, to such an extent that he lost his job and was threatened with divorce. He had become totally indifferent to all that concerned him. The psychiatrist who was following him for depression noticed signs of raised intracranial pressure. A meningioma, 6cm in diameter, situated under the midline of the frontal lobes was diagnosed by MRI. The surgery, performed 30 years after Elliot's, was without sequelae. Etienne recovered his normal behavior, went into practice, and recovered his family life. Meningioma now has a wholly positive prognosis.

Chapter 9
Altered States of Consciousness

In 1975 a book appeared which was to have widespread repercussions: *Life after Life*, by American psychiatrist Raymond Moody. It hit the best-sellers list not least because of its preface signed by Elisabeth Kübler-Ross, a physician famous for developing a method of terminal care of the dying. From her clinical experience, she recognized cases of "survival" after apparent death as commonplace.

Since then there has been a proliferation of literature exploring this subject that naturally excites the curiosity of the general public. Is there proof of a life after death? Or, can consciousness exist outside the brain? These problems relate to the temptation to accept dualism, alluded to earlier in the present book, but subtly disguised: the aim is to reveal the existence of an invisible spirit and find its proof in the visible, the measurable, in our practical experience.

These altered states of consciousness must certainly exist, for a large number of case studies demonstrate their universality, but the conclusions that are drawn are often manifestations of their authors' prejudices: they seek support for their beliefs in a very partial interpretation of such experiences. It should be emphasized that experiences on the border between life and death are not accounts of the dead who rise again and tell what has happened. The "next world" is not linked to our own by a sort of bridge that it is possible to cross in both directions. Real death is irreversible and no one has come back to life.

An out-of-body experience (OBE) refers to the phenomenon of existing outside one's body, whereas a near-death experience (NDE) relates what happens as death approaches. The two are not directly related but can happen successively in certain cases and certain circumstances.

To cite Moody's classic description of an OBE followed by an NDE:

A man is dying and, as he reaches the point of greatest physical distress, he hears himself pronounced dead by his doctor. He begins to hear an uncomfortable noise, a loud ringing or buzzing, and at the same time feels himself moving very rapidly through a long dark tunnel. After this, he suddenly finds himself

J. Neirynck, *Your Brain and Your Self: What You Need to Know*,
© Springer-Verlag Berlin Heidelberg 2009

outside of his own physical body, but still in the immediate physical environment, and he sees his own body from a distance, as though he is a spectator. He watches the resuscitation attempt from his unusual vantage point and is in a state of emotional upheaval.

After a while, he collects himself and becomes more accustomed to his odd condition. He notices that he still has a "body," but one of a very different nature and with very different powers from the physical body he has left behind. Soon other things begin to happen. Others come to meet and to help him. He glimpses the spirits of relatives and friends who have already died, and a loving, warm spirit of a kind he has never encountered before – a being of light – appears before him. This being asks him a question, non-verbally, to make him evaluate his life and helps him along by showing him a panoramic, instantaneous playback of the major events of his life. At some point he finds himself approaching some sort of barrier or border, apparently representing the limit between earthly life and the next life. Yet, he finds that he must go back to the earth, that the time for his death has not yet come. At this point he resists, for by now he is taken up with his experiences in the afterlife and does not want to return. He is overwhelmed by intense feelings of joy, love and peace. Despite his attitude, though, he somehow reunites with his physical body and lives.

Later he tries to tell others but he has trouble doing so. In the first place, he can find no human words adequate to describe these unearthly episodes. He also finds that others scoff, so he stops telling other people. Still, the experience affects his life profoundly, especially his views about death and its relationship to life.

Neurology of the OBE

Three types of OBE have been described:

- the autoscopic hallucination, when a subject sees his double in the space around him without actually leaving his own body
- the true OBE, when the double observes the real body
- the heautoscopy when the subject sees his double without being able to decide if he exists in his physical body or in the double, or even if he exists in both at the same time by a process of bilocation.

In all three cases the subject sees himself, hence the term autoscopy. But it is only in the autoscopic hallucination that the subject understands the illusory nature of the phenomenon, whereas the true OBE and heautoscopy are experienced as completely realistic events.

These phenomena are very striking for a subject because they defy the perceived sentiment of the unity of the person and the body, of the real me in my body, which is the subject of both one's experience and one's activity. Studies by Olaf Blanke at the EPFL demonstrate that the determining role in the triggering of the OBE depends on a restricted area of the temporo-parietal cortex, comprising the angular gyrus and the superior temporal gyrus. This discovery was made by chance while investigating epileptic patients for surgery by recording and stimulating their cerebral cortex. A hundred or so electrodes were inserted in the right hemisphere. When one of them was stimulated the patient experienced an OBE.

Why is this cortical area involved? Perhaps because it is adjacent to the vestibular cortex which is involved in our perception of the body in space, and also to the association cortex that integrates tactile, proprioceptive and visual sensations of our body. The impression of a dissociation between my self and my body may be a result of an incapacity to integrate vestibular and sensory information.

A more detailed study would be useful in providing a better understanding of the mechanisms of our usual experience that our body and our consciousness of it are collocated. In the end, what is astonishing is not that we experience leaving our body behind, but rather that, normally, we feel that we are indeed inside our own body. The OBE occurs in about 10% of the population, independent of cultural background, in the face of various pathological conditions such as meningitis, encephalitis, poisoning, epilepsy, migraine and head injury, but also during sleep, general anesthesia, free fall, tiredness and anxiety states. Neither hemisphere seems dominant in this phenomenon, except that autoscopic hallucinations and OBE tend to be associated with lesions of the right hemisphere and heautoscopy with the left. Whatever, these phenomena only occur in cases of disturbance of visual and vestibular function, both of which help us to orientate in space. The cortical areas related to these phenomena can be stimulated electrically to artificially trigger them. So our present understanding of them is that they represent a failure to properly integrate a number of different perceptions of our body by our brain.

This explanation is of course rather disappointing for supporters of dualism. Various people have attempted experiments to try to prove that there is, after all, a true disembodiment and that a spirit with real sensations truly escapees from the body. In England, Peter Fenwick, a specialist in neurological psychiatry, fixed on the ceiling of the emergency room a number composed of several digits. About 10% of reanimated patients reported disembodiment. By interviewing them later, Fenwick tried to ascertain how many had seen the number. The result is amazing: patients reported that they noticed the number on the ceiling but that they had much more interesting things to do than read it. Sylvie Déthiollaz of the Noêsis Center in Geneva has conducted similar experiments using a subject who volunteered to attempt disembodiment, and a series of geometric images stored on a computer. One of the images was generated randomly by the computer in a different room from the subject. On a significant number of occasions, the subject managed to identify the image, although it was out of his sight and even the experimenter did not know which image was on the computer screen. But this result with a single subject can in no way be considered as conclusive, and more work is in progress.

Neurology of NDE

In December 2001 an article appeared in *The Lancet* reporting that 18% of patients resuscitated after a cardiac arrest described NDE and that 12% had a more or less clear recollection of the phenomenon. Moody's description, cited above, is of a positive experience. He neglects negative recollections such as being tortured by demons, that are reported by 15% of patients.

Usually NDE is associated with states of the brain brought on by respiratory or cardiac arrest, that is say oxygen starvation of neurons. However, NDE can be produced in patients that are not in mortal danger, but rather are suffering from psychoses due to deficient neuronal transmission or drug intoxication, for example with marijuana or LSD. Similar experiences can be triggered by electrical stimulation of the temporal lobe or hippocampus, or by lowered oxygen levels in the blood. It has to be said, however, that NDE evoked by stimulating the brain is not typified by the sensation of serenity that is a characteristic of much true NDE. This latter effect of relaxation and peace may well be due to the secretion of endorphins by a patient under stress.

Divergent Opinions

According to Moody, OBE and NDE prove the existence of a soul, distinct from the body, that promises eternal life, a sample of which is perceived at the moment of death. The detailed revelation of one's lifetime experiences is held to be a judgment that results in the opening, or not, of the gates of heaven. For Moody, a convinced optimist, everyone goes to heaven. For others, negative NDE supports the existence of hell.

The story of the woman who spoke in numbers

Séraphine was an inmate of a retirement home near Toulon, and was visited regularly by her family doctor. One morning he found her fit and well, in high spirits and freshly made up. The following dialog ensued:

- So my dear, how are you?
- Twelve times twelve equals 144.
- Did you sleep well?
- Nine plus eight equals 17.
- Your room is not too warm?
- Nine times seven equals 63.

The conversation continued in a similar fashion for over half an hour. The doctor prescribed an intravenous vasodilator and by the evening the episode

had come to an end and Séraphine was speaking normally. What had happened was a vascular spasm, a transient ischemic attack (TIA). The spasm had not affected the speech area of the brain for Séraphine could speak normally, and had no difficulty finding her words. Only her vocabulary was affected, while her mental arithmetic was not only intact but enhanced. Séraphine had worked all her life at the check-out of a grocery store.

Chapter 10
The Myth of the Artificial Brain

For a long time, engineers were no more interested in consciousness than philosophers were interested in the steam engine. The typical engineer built supposedly neutral machinery: its ultimate usage would depend on ethical choices based on contemporary views in another domain, that of thought. The techniques of the time related to energy, materials and buildings. A stove, a house, a bicycle, a hammer, all were extensions of the human body, helping to build a better life, but not to think. In a dualist culture, science and technology deal exclusively with matter, while religion and philosophy deal with the soul.

This essentially materialistic technology was neutral in itself. It was judged on the basis of Platonic ideals, rooted in a rigorous transcendentalism, without apparent reference to the society that nurtured them. The developed world lives, and often continues to live, in this dualist scenario, without even realizing it.

We are slowly abandoning this dichotomy, where those that deal with matter leave others to deal with the soul. By inventing the concept of software, information technologists have given material form to the workings of the soul or, on the contrary, breathed a soul into matter. They are more and more obliged to analyze behavior and the function of the brain if they claim to be able to decode speech or compress images. After inventing the pump by modeling it on the heart and the bellows on the lungs, after dreaming up spectacles and the artificial kidney, hearing aids and artificial knees, after providing a host of prostheses for the human body in order to extend its capacity or repair its defects, engineers have now reached the last of the organs that still defy them, the brain.

As we saw earlier, engineers assist the medical profession, offering all sorts of ways to investigate the brain: EEG and MEG, MRI and PET. Apart from their use in fully-fledged medical practice, these techniques enable the elaboration of models of the brain by analyzing, with greater and greater precision, the activity of its different parts, according to the functions they serve.

At this stage we can ask two questions. What prostheses can we suggest to surpass the performance of the brain or to offset its deficiencies? Does the brain possess functions that cannot be automated? And, underlying these two questions a fundamental problem recurs: can our whole intellectual activity be explained in terms of brain function? Dualists will unequivocally answer no; neurologists will utter a guarded yes. When a machine appears that is indistinguishable from a man, the problem will be solved.

J. Neirynck, *Your Brain and Your Self: What You Need to Know*,
© Springer-Verlag Berlin Heidelberg 2009

The Hesitant Birth of the Computer

The invention of writing 5000 years ago constitutes a first advance toward information technology. It allowed us to stock data or facts, maybe the accountancy of a farm or the chronicle of a reign, that would otherwise slowly disappear from human memory. The first writing consisted of ideograms each representing a word. An important step was taken with the invention of phonetic script around 3000 years ago. The Phoenician alphabet constituted a quite remarkable advance in codifying more than a hundred phonemes, used in a variety of languages, by 26 letters. The invention of printing during the Renaissance multiplied the power of writing.

The first modern techniques in information technology, the telegraph of 1842, the telephone of 1846 and the radio of 1896, limited themselves to transmitting information through space without modifying it. They were interesting but not surprising. When the first computers appeared in the 1940s, certain enthusiastic commentators thought it fitting to baptize them "electronic brains". At that time it seemed quite extraordinary that a machine could perform a long series of calculations with great precision and obtain results that were equal to or better than those of the human brain. Spirits were equally elated when in 1997 a computer named Deeper Blue beat Garry Kasparov at chess by 3.5 points to 2.5 However, programmers had taken 40 years to reach this performance and they used an IBM machine weighing 700kg against a brain weighing 1.5kg.

The motivation and the financing for the first prototypes during the Second World War revolved around an objective that we now know to be unachievable on theoretical and practical grounds, that is long term weather forecasting. One concrete objective was to improve forecasts at the time of the Normandy landings in 1944, that were threatened by disaster in the event of bad weather. People imagined at the time that it would suffice to increase the number of meteorological stations and to have access to enough calculating power to solve a large number of well-known equations involving physical parameters such as pressure, wind speed and humidity. However, no-one did better than forecasts over a few days.

We had to await the 1960s for the discovery of the concept of chaos which annihilated any such hope: meteorology is a deterministic phenomenon, but it is unpredictable because it depends on initial conditions that are not known with infinite precision. We understand the equations that describe the atmosphere, but we can never solve them sufficiently accurately, however many measurements we make or however powerful our computers. The most powerful computer conceivable will never resolve a poorly formulated mathematical problem.

This historical glimpse is interesting because it demonstrates how engineers can, by tackling insoluble problems, resolve other problems that they had never envisaged. Certainly, the computer does not permit accurate long term weather forecasting, but it has proved an indispensable office machine, essentially replacing the typewriter, the calculator, the tabulator and the drawing board.

In fact, once the first enthusiasm was past, it was realized that a standard computer does no more than compress the time taken for the execution of a program entirely conceived by a human brain. Its workings are entirely predictable: it is no more nor less than a preprogrammed robot, like the automatons of the eighteenth century that were based on mechanical technology. With the change to electronic technology, the robot's performance improved quantitatively, but not qualitatively. Its enormous speed of calculation disguises its simplistic conception.

Two Decisive Inventions

The present success of the standard computer is due to a large extent to two inventions.

The first is the integrated circuit, on a minute silicon chip, which allows the very fast handing of information at a ridiculously low cost and small size, and with modest energy consumption and, therefore, acceptable heat generation. Over the last generation, the average density of the circuit components doubles every two years, and will continue to do so for a number of years. A chip integrates tens of thousands of components, and even millions, on a surface of a few square millimeters. It has replaced the cabinets stuffed with vacuum tubes in the 1950s or with transistors in the 1960s. The space occupied, the energy consumed, and the high initial cost limited computers at that time to specific tasks involving scientific calculations that were too complicated to be undertaken manually.

If electronic components get smaller, it means that information crosses them faster so that more operations can be performed in the same time. A chip containing a large number of components can be manufactured in large numbers, thus reducing the unit cost. Lower costs mean more potential customers. Such large-scale economies add up to a surprising result: the relative cost of an operation on newer computers that come into service with ever increasing frequency is cut in half every three years. The same operation in 2000 cost ten thousand times less that in 1960 on an IBM 650 machine. To sum up, one may say that the computer, which was conceived by an English mathematician, Charles Babbage (1791–1871), in the nineteenth century using mechanical technology, and adapted by another mathematician, John von Neumann (1903–1957), working at Princeton in the 1940s, only began to realize its full potential thirty years later with the development of components suited to the task.

The second invention was software that allowed computers to be operated by users who understood nothing about how they functioned, but who could then access them with quite elementary and intuitive actions via a keyboard and mouse. The invention of the first high-level language, Fortran, dates from 1954, and since the 1980s numerous specialized programs for word processing, graphics, accounting and management have simplified the widespread use of an instrument of which the origin was in a radically different context. In other words, the computer invented by cutting-edge scientists for their own exclusive use has ended up in the public domain thanks to an invention that completely masks the incredible complexity of the operations that take place within the black box.

This remark will throw some light on what is to follow. The very intelligent and convivial behavior of modern personal computers depends on the activity of a multitude of elementary components carrying out rudimentary operations. The computer user operates an intuitive model of a machine that can act as a secretary, or as a partner in a game. He expects a certain degree of intelligence from a computer, forgetting that it is a mere programmed robot. He does not try to master it by studying the instructions or by deciphering the software: indeed this is almost impossible, and practically useless. He is annoyed when his computer hangs up or gives unexpected responses. The user personalizes the machine before him, and finishes by attributing it with human characteristics: it is tired, it needs to rest, it plays up.

The Two Limits of von Neumann's Computer

Let us reflect on the characteristics of the classic computer according to the concept of John von Neumann. This involves:

- a central processor, perhaps with some tens of duplicated units
- the serial handling of information by this processor
- a program stored in a memory which indicates the series of operations to be undertaken by the processor
- binary coding of information, that represents numbers, letters, images or sounds by sequences of the numbers 0 and 1
- an addressable memory, access to which does not depend on content but on the place where the information is stored.

Such a computer is just one machine for processing information among many others that one could imagine. The historic origin of this concept and its good performance in certain domains should not make us forget that there could be others, with even better performance, for other applications.

In fact, digital computers that have spread throughout many sectors of activity excel in certain domains precisely because they do well what the human brain does badly. Repetitive tasks, such as keeping files up to date, correcting spelling mistakes in a text, or the multiplication of two numbers of multiple digits, lead to numerous errors when they are performed by a human brain, which is quickly overloaded, or fatigued by work for which it was not designed and on which it has difficulty concentrating. Modern computers are prostheses which extend the activity of the brain beyond its natural limits in specific directions, but not in others. Similarly, a car allows its driver to travel twenty times faster than he could on foot, but it does not allow him to jump over a simple fence. In practical terms the most fruitful application for computers at present is in administration and accounting, much more than scientific calculation, for which they were originally developed, but which offers limited perspectives for development and profit.

Nevertheless we can easily recognize two categories of task that the human brain performs better than the most sophisticated digital computer, even when we multiply the number of its processors and make them work in parallel, or if we equip our computer with an expert system, that is feeding it with experience derived from human brains. These two adjuncts to the initial design represent attempts to extend the field of application of the standard computer without interfering with its basic structure.

Voice Processing

The first category contains tasks that have two characteristics in common. First, tasks for which it is not possible to provide a precise definition so as to be able to conceive an algorithm, that is to say a series of mathematical operations that will produce a result. Second, tasks that imply the **approximate** processing in real time of a mass of data, for which the human brain is especially adapted.

A typical example of such tasks is processing of the voice. For the moment we are still obliged to drive our computers by mechanical actions, for example with a keyboard and a mouse. It would be so much simpler to be able to use our voice to manipulate a typewriter, a telephone or an airplane. For the first example, it means inventing a dictating machine that would type a text correctly on the basis of a human voice. In 1971 such a project was launched in the United States, called ARPA-SUR (Advanced Research Projects Agency - Speech Understanding Research).The aim was to analyze a conversation containing a vocabulary of a thousand words, regardless of who was speaking. The best performance around that time was obtained by a laboratory at Carnegie-Mellon University which produced a prototype using a standard computer that worked for a maximum of five voices.

Thirty years later performance has improved, but the return has not matched the amount of work invested by numerous teams. We can now obtain commercial software that will recognize, with 95% reliability, isolated words from a vocabulary of two or three hundred words as pronounced by any speaker. Its cost of a thousand dollars or so excludes its widespread application, in using a telephone for example. In addition, dictating machines exist that will recognize a single speaker after a preliminary training session: these machines must learn how a given speaker pronounces the words. Their performance is noticeably degraded when the voice recognition takes place in the face of background noise or when the user has a marked accent. Also, the decoding operates with a slight delay even for short phrases. There exist prototypes in research laboratories that perform better, but their high price discourages their mass commercialization.

To summarize, the brain of a child of three performs better than any machine so far in the elementary task of recognizing a voice.

The difficulty of the problem is its very imprecision. Pronunciation of phonemes is very variable depending on individual speakers, and isolated phonemes cannot be identified with certainty. In reality, the human brain recognizes strings of phonemes, using its own data base of familiar words. It also depends on the whole context of the verbal message. If not, it would be impossible for the auditory message to be distinguishable. The context is really important to distinguish, for example: *hear* and *here*, *talk* and *torque*, *ice cream* and *I scream*.

A computer would solve these problems by using algorithms that would take so long to execute that it would be impossible to use them in real time. The task would become unbearable when, instead of pronouncing the words carefully separated from each other, the computer were exposed to real, everyday language with all its liaisons, elisions and imprecision.

This example of voice recognition demonstrates that a human brain succeeds better than a machine in this context to such a degree that one must doubt the possibility of any real success of such an enterprise. We find ourselves confronted by a problem that doubtless has a simpler technical solution, as long as we distance ourselves from von Neumann's computer, and we stop assuming that this model is good enough for everything.

The same comments can be applied to problems of image recognition by the brain that are hard to undertake with the methodology of classic calculation. How do we tell a computer the difference between a tree, a telephone pole, and a feather duster?

Finally, automatic translation of a text is still in deadlock today in spite of half a century of effort. In 1954 IBM demonstrated the possibility of translating fifty or so phrases from Russian to English. Orders, especially military ones, poured into research laboratories. But in 1966 a first assessment by a special agency of the American government acknowledged that translation by humans was of a much better quality for a lower price. It has been estimated that today automatic translation is used in of the order of a thousandth of the global translation market, that is estimated at 10 billion dollars and of which 99.9% is achieved by human brains that are well-practiced in the job.

Practically Insoluble Mathematical Problems

There is a second category of tasks that is beyond the scope of the standard computer. This time it involves purely mathematical problems that have been well studied theoretically, for the solution of which the algorithm is known and programmable, but which would take too long to solve with an ordinary computer. Such problems are those of the "commercial traveler", "map coloring", the Steiner tree, and the Hamilton and Kirkman problems, and many others. The first of these consists of a commercial traveler visiting all his customers using the shortest possible route. The second involves the coloring of, for example, a map using four colors such that no two neighboring countries have the same color.

For all these problems only so-called non-polynomial algorithms are known, implying that on a standard computer the time needed to process them increases explosively and exponentially with the dimension of the problem to the point that they are practically insoluble. For example the processing time is proportional to the nth power of 2 when n is the dimension. The visit of the commercial traveler to five towns can be tackled by a computer, but it is overwhelmed if we pose the same problem for 30 towns. The time taken for the computer to solve the second problem is a billion times longer than for the first! In contrast, when confronted with such problems, the human brain quickly comes up with suggested solutions, which might not be the best ones, but which have the merit of being practical and quick. In everyday life we do not necessarily want the best solution, for we do not have time to wait.

When faced with these two categories of problems, the von Neumann computer fails because it tries to find:

- the optimal solution
- by serially processing
- one after the other
- a mass of hardly significant data.

The vast majority of the bytes handled by a digital computer have strictly no significance, for the apparent precision of the calculations is disproportionate compared to the limited precision of the data and the equally limited requirements of the results. The machine's efficiency, and its precision, are much less than are generally believed. It is better to find an acceptable solution in a limited time rather than waste time trying to find the ideal solution.

To sum up, von Neumann's computer is an ideal tool to solve all problems that can be solved by an algorithm in a finite time. It is therefore basically inadapted to resolve two categories of problem: those for which we do not know an algorithm, and those for which an algorithm is known but which would need a disproportionate amount of time to calculate. It beats the brain where the latter is weak, but it is beaten everywhere else.

Comparison Between Computer and Brain

Von Neumann's computer is a machine for processing information that is built rather differently from the brain:

	Computer	Brain
Number of processors	between 1 and 100	100 billion
Information coding	digital	analog
Control	by program	by learning
Timing	nanoseconds	milliseconds
Solidity	weak	strong
Results	precise	approximate
Memory	addressable	associative
Nature of the processors	common	specific

It was understandable that the von Neumann computer was developed first, for it was a radically different machine from the brain and did tasks that the brain did badly. But there is no reason to consider this machine universal and not develop others according to different models more adapted to a given function. Further, there is no reason to attribute a computer with properties that are particular to the brain such as creativity or consciousness, as was done naively in the early days. Recent developments, such as expert systems or multiprocessors, fuzzy logic or infographics, can seem to emulate the brain as they extend the use of the computer closer and closer to human cerebral function, but such present or future developments do not change its intrinsic nature of an inherently predictable preprogrammed robot.

Artificial Neuronal Networks

In 1943 an American neurophysiologist Warren McCulloch and logician Walter Pitts proposed a very simple model of a biological neuron in the form of a circuit that performed additions of signal inputs, each of which was weighted by a coefficient, and which produced an output signal if the sum exceeded a certain threshold. These authors demonstrated that a system of artificial neurons of this type, functioning at a synchronous rhythm, could perform the same sort of calculations as a digital computer basically composed of logical circuits that perform the well-known operations "AND" and "OR". The output is "1" when two entry signals are "1" in the first case, and the output is "1" if one entry signal is "1" in the second case.

The spectacular success of von Neumann's computer totally eclipsed the theoretical possibility opened by the work of McCulloch and Pitts. What is more, engineers knew through experience that it was not always fruitful to copy nature rigidly. Cars do not walk on legs and airplanes do not beat their wings. The idea of simulating brain function scrupulously did not seem to offer much interest in itself at the time. But half a century later the limitations of the classic computer have rekindled interest in other types of computer based more closely on the human brain.

We had to await the1960s and 1970s for a few isolated researchers to begin to build and use artificial neurons. The names of Widrow, Grossberg, Kohonen, Rosenblatt and Sejnowski marked that period. But it was a paper by American physicist John Hopfield on neural networks in1982 that attracted the attention of the technical community to this so far poorly explored field. Since then we have witnessed a veritable explosion in the domain, as can be judged by a number of recent articles and reviews. Several scientific journals on neural networks and "neurocomputing" have been created aimed exclusively at this subject. Conferences and seminars abound. Researchers have turned their attention and ambitions toward until recently largely unexplored fields. One might refer to a colloquium on automatic face and gesture recognition held in Zurich in 1995 which presented 60 contributions on a subject that one might consider as very narrow.

Numerous applications constitute subjects of research, including recognition of voice, writing and image, synthesis of voice from written text, problems of optimization, control and robotics, associative memory, medical diagnosis and monitoring, modeling of chemical and physical processes, and the conception of telecommunication networks.

All these applications have in common that they use very simple networks comprising a limited number of artificial neurons. The synapses, in this case represented by the weighting in the calculations made by the artificial neuron, are usually not programmed, but established by learning, that is to say after exposure to a number of examples. The comparison of the desired results with the result obtained by the network leads to modification of the weighting, positively or negatively according to the comparison.

It is important to emphasize some nuances between biological and artificial neurons. These differences are often not considered to be pertinent, but there exist a few examples of applications that suggest that the definition of the basic neuron could be decisive in the functioning of the network.

- Firstly, we see a tendency to remain faithful to the well-known domain of binary logic, clinging to a threshold paradigm, that is to say one that operates in an all-or-none fashion. As long as we remain dependent on binary logic, we distance ourselves from the biological neuron, that has a continuous, non-linear response. It might be of interest to develop a frankly analogical machine, for the brain has to solve tasks that are not at all binary.
- The biological neuron responds with a series of impulses of which the frequency expresses the level of excitation. It is not obvious that this is the same as the continuous signal from an artificial neuron of which the amplitude is proportional to the impulse frequency. Perhaps the phase of a biological signal, that is the interval between the impulses, is also significant for the human brain.
- Biological neurons are not locked to a synchronous clock. That is the case for most circuits of artificial neurons that are thus based on conceptions from classic computer technology.
- Biological neurons are not completely deterministic systems. Under the same conditions of excitation their output signal varies according to random rules. This could be, and should be, incorporated into artificial models.
- Finally, biological neurons are not simple systems to add input signals, as we saw in Chapter II. The role of glial cells should probably not be disregarded either, and they are completely absent from networks of artificial neurons.

The electronic realization of the model of McCulloch and Pitts or its numerous variants is not in principle a problem. It consists of a circuit that calculates a weighted sum of input signals that can be constructed with an amplifier and a few resistances. It becomes complicated when we have to provide resistances of which the value varies with learning, and when we have to connect the output of an artificial neuron, the equivalent of its axon, with ten thousand inputs to other neurons, via the equivalent of their dendrites. Classic integrated circuit technology is not adapted to this sort of requirement.

An original field of research has developed as a consequence, that of cellular neuronal networks (CNN), that use the principle of extended networks of artificial neurons that have connections limited to their immediate neighbors. Theoretical studies of this question demonstrate that such networks are well adapted to certain problems of recognition. This projects merits attention: it is not indispensable to slavishly copy the brain to reproduce some of its functions. In view of the multitude of parameters available, a very limited degree of connectivity can be compensated by increased speed, for example.

At the present time, research into artificial neurons, and even their applications, uses classic computers. They allow the simulation of the behavior of a network of a few tens of neurons. But there is an obvious contradiction in using a machine that is basically unable to resolve certain problems to simulate another machine that should be able to resolve them. In so far as the neuronal network is expected to perform tasks, such as voice recognition for example, that a classic computer cannot perform, the latter cannot be expected to simulate the process. In the end, many so-called neuronal networks simulated on a classic computer are only a means, that is attractive to engineers, of programming statistical analytical paradigms. A competent mathematician could tackle them directly without the almost devotional evocation of the neuronal network.

In other words these "neuronal networks" have been used to resolve problems for which they were not essential, with people imagining that they would be able to extrapolate solutions found by simulation on a classic computer to problems of real neuronal networks, when the dimensions of the problem in hand made it really necessary. But the question remains: is there not a qualitative leap between the behavior of a few tens of neurons and that of a few million? Will methods that are adequate in the first case also be applicable to the second? The human brain would not have evolved to have 100 billion neurons if 100 thousand would have sufficed. So the present developments in artificial neuronal networks must face this paradox, and find themselves in a deadlock. On the one hand a host of problems exists that cannot be solved by the classic computer; on the other, it might in principle be possible to construct machines to reproduce certain functions of the brain, especially for shape recognition. But it is not possible to conceive a design for these machines using the von Neumann computer, that is at the moment the universal savior of engineers stuck for a theory.

Today's essential contribution should be a usable theory. But interesting neuronal circuits seem systematically to have a non-linear behavior, and the theoretical basis of such circuits remains very fragmentary. At best they provide qualitative results, for example to answer the question as to whether there exists one, several or no stable rest positions in a network. But it cannot tell us the number, or form, or extent of the basin of attraction of each of them.

Nevertheless the real challenge is substantial: how can we conceive and construct a machine of significant size, that is to say with thousands or tens of thousands of neurons, using integrated circuit technology, but with no possibility of predicting the machine's behavior? How can we resolve realistic, and therefore complex, problems neither knowing what an adequate machine would consist of, nor what would be a good learning algorithm? In view of the mass of choices that confront us, there is little hope of finding the right solution by chance. There is an enormous risk in tackling poorly understood practical problems using a machine that we really do not know how to build.

The list of topical open questions in the domain of artificial neuronal networks is long.

- Problems of architecture: how many neurons do we need for a given perform-ance? Should we organize them in layers or otherwise? How many input and output connections should we envisage for each neuron? What relationship between inputs and outputs are necessary? Should the neurons' activity be syn-chronous or asynchronous, deterministic or random? Should we plan a second, parallel system to imitate the performance of glia?
- Problems of programming: Should the connections between neurons be prede-termined (innate) or learned (acquired)? If learning is involved, how many trials are necessary before a skill is established? Should the learning phase be separate from that of the acquired function or should learning be continuous during functioning?
- Problems of function: What is the relationship between architecture and task? What multiple tasks can be performed with a given architecture? What is the sensitivity of a given neuron to a failure or the presentation of imprecise data?

These are some of the enigmas that we must resolve if artificial neuronal networks are to emerge from the laboratory and replace von Neumann's computer in those situations where it does not have its place. If decisive breakthroughs are not achieved soon the whole domain is threatened with disappearance through a fatal combination of disappointing performances in the face of unreasonable expecta-tions. Whole cohorts of researchers are devoted to this domain. The stakes are huge, for the computer, as we know it, could be remembered in a few decades as a simple prototype, limited to functions that are hardly different from those of a typewriter, a calculator, or a mere accounting machine, all of which appeared at the beginning of the twentieth century and have disappeared since.

Braitenberg's Project: Mechanical and Evolutionary Psychology

Everything so far depends on the most classic forms of technology. Engineers have worked hard to compensate for the defects of the von Neumann computer by creat-ing special processors for special functions, particularly in terms of the recognition of shapes. At this stage we can describe this approach as trying to diversify compu-ter science by abandoning von Neumann's prototype, that has for too long enjoyed a monopoly that it does not merit. Further, recent work demonstrates that, by successfully simulating certain functions of the brain, engineers can aspire to more ambitious goals: in the end, would it not be possible to create a universal machine that could fulfill all the functions of the brain?

The motivation, whether conscious or not, that animates the current fashion for this type of research is essentially methodological, even philosophical. Is it not a reflection of the ambition to settle the fundamental debate, that we are discussing here since the beginning of this book, between materialists and dualists on the existence, or not, of a spirit or soul, an immaterial entity that can influence the brain?

The novelty of artificial neurons is methodological. Instead of scrutinizing the findings of psychology or experimental neuroscience, of interpreting and analyzing them to throw light on this philosophical question, would it not be more expeditious to construct a robot with all the functions of human intelligence, including sensitivity and creativity. Would we so demonstrate experimentally that man, even in his scientific, esthetic and affective activities, is in the end nothing other than an automaton, entirely conditioned by his genetic heritage and his existential apprenticeship? If this were true, human liberty would be a mere illusion; man would be determined in the same way as any material object in the universe, neither free nor responsible for his acts.

This challenge of synthetic psychology has been taken up by German neuroscientist Valentino Braitenberg in his ground-breaking research, that is both serious and amusing, and that is summed up in his book "Vehicles, Experiments in Synthetic Psychology" published in 1986. Without submerging himself in technical detail and without slavishly copying existing neurons, he created, in the sense of a demiurge, fourteen generations of "vehicles", little robots with wheels, each generation showing an essential qualitative difference compared with the previous. By beginning with simulating primary reflexes, he finally achieved apparent manifestations of higher sentiments. From expressions of fear, aggression and logic, the results progressed to foresight, egotism and optimism. He only needed a few electronic components to realize a stupefying plethora of reactions. It was not even necessary to actually build these vehicles to demonstrate them: any competent technician can imagine them and admit the plausibility of their behavior.

Braitenberg systematically copied the mechanisms of biological evolution to achieve a virtual reproduction of its activity using electromechanical rather than biochemical building blocks. There was no question of deliberately conceiving very sophisticated machines for a precise purpose: it is easier for a creator to let machines breed themselves randomly and provide a mechanism for the elimination of the least well adapted ones. So, we might randomly construct a few dozen prototypes that roam around a table, and eliminate any that fall off. Without prior planning we shall have discovered how to construct a robot capable of distinguishing a flat surface from its edges.

It will be obvious that this type of research remains strictly in the context of behaviorism: the only things that count are objectively observable behavioral phenomena. Behavior that is analogous to that of highly evolved animals, and even man, is obtained by connecting a handful of electronic components. We are reminded here of the robots that were actually built in the 1960s and 1970s as demonstration models, such as a mechanical mouse that could learn to navigate a maze. However, the interest of such an approach is limited and Braitenberg's work is more of a series of essays, somewhere between literature and philosophy, than a really scientific contribution. In summary, his book is an interesting document for it demystifies the pompous discourses of certain psychologists and it astutely supports the materialist, evolutionist hypothesis.

The Blue Brain Project

The Blue Brain project at the EPFL runs on a Blue Gene computer conceived by IBM for the EPFL. It benefits from a speed of 22.9 teraflops, which means in simple terms that it can theoretically perform 22,900 billion operations per second. It contains 8000 independent processors, compared with one in a personal computer. This puts the Blue Brain computer among the fifteen fastest in the world. In comparison, the Blue Gene at the Lawrence Livermore Laboratory at Berkeley achieves 367 teraflops.

The project leader is Henry Markram whose aim is to simulate a cortical column, a fundamental component of the cerebral cortex. Such a tiny column, half a millimeter in diameter and two or three millimeters high contains from ten to a hundred thousand neurons and five kilometers of "wiring". To begin with, Blue Brain is attempting to simulate a column in the rat cortex. The method is to analyze the behavior of each neuron physiologically and its connections morphologically, and then to write equations incorporating these data. The cortical column can be considered to represent emerging intelligence in mammals, for there are none in reptiles. When evolution produced columns in its random progress, it immediately adopted them because they proved useful for survival. For the biological brain they represent the equivalent of what microprocessors represent for the digital computer. In its heart exists the secret of all that man imagines, invents and feels.

As a second stage, the aim is to multiply cortical columns in order to reconstruct the equivalent of a digital brain. Theoretically this will allow some animal experimentation to be bypassed, for voluntarily connecting or disconnecting each neuron with simple computing operations should, at least partly, reproduce the equivalent of lesions in a living brain. One of the major interests of the Markram team is cerebral plasticity. So the idea is also to follow the modifications that the brain undergoes spontaneously during its continuous reorganization, reinforcing certain synapses and suppressing others. Memory is not locked in inert structures, but in ones that are constantly modified. Our capacity to process information, whether for seeing, hearing, walking, reading, or whatever, is acquired by apprenticeship, repeating the operation until it becomes unconscious.

It this way we may ultimately be able to investigate, not *in vivo* or *in vitro*, but *in silicio*, the roots of different brain diseases and study the effects of medication. Instead of depending on laboratory experiments that take months or years, the same work would thus be accomplished in a few days.

But for the moment only one column can be simulated (and then not all its functions or parameters). The study of the whole brain will depend on the availability of machines a million times more powerful. We shall simply have to wait, but computers are following a rapid growth curve that has seen their power develop in the last half century by an equally large factor. IBM expects Blue Gene/P to increase its speed by a thousand times in the next ten years.

The Blue Brain project marks a return to von Neumann's computer to simulate an organ, the cortical column, that in fact processes information according to totally different principles. As much as the project may prove useful to understand the detailed functioning of one part of the human brain, it seems inappropriate as a means for its simulation. It is a path that is worth exploring, without however expecting it to provide any more than a convenient model for the abstract study of cerebral function.

Bio-Inspired Computing

Daniel Mange at the EPFL was an initiator of an approach to computing based on biology. His "BioCube" is a prototype among others of what systems computing may use in a century from now. By then the computer will perhaps be a network of processors on a nanotechnology scale, that is to say with dimensions of the order of a millionth of a millimeter. A prototype produced at the EPFL contains 4×4×4 = 64 nodes. Each node contains a processor constructed with the help of the most recent programmable logical circuits. Each is in direct contact with its six immediate neighbors and calculates its future state as a function of the present states of these. It is a form of the cellular neuronal network (CNN) mentioned earlier: without slavishly copying the function of the biological neuron, we are resolutely distancing ourselves from von Neumann's computer.

Once it is reduced to the microscopic dimensions of a network of molecules, the nanocomputer will be characterized by properties belonging to living organisms: self-reparability, robustness, and reliability. It will be capable of growth by cellular division and differentiation, and of reproduction. By introducing a random factor in this process it is possible to envisage generations of bio-inspired computers engaged in a creative revolution. We shall rediscover the evolutionary process imagined by Braitenberg.

Creation of Consciousness

Braitenberg's vehicles, Markram's Blue Brain and Mange's BioCube all have the merit of posing a major question: just how far can we go in studying man's highest faculties by constructing machines? Supposing that we succeed in a detailed autopsy of a brain going as far as a census of all the synapses, and that we reconstruct this brain using artificial neurons, will it still have the same original function as its biological prototype? Will it be able to speak, write, recall memories and show emotion?

Nobody doubts that one's personality is preserved even if certain body parts, such as arms and legs, teeth, kidneys or heart, are replaced by artificial prostheses or transplanted organs, as long as the brain is not involved. But what would happen to personality if the whole or part of the brain was replaced by a prosthesis or transplant? Science fiction has not infrequently shown an interest in the hypothesis of a man with a brain transplanted from another man or with an artificial brain. This is not a present-day concern of physicians or engineers, for they are too realistic to bother to evoke a project that is beyond their grasp, and will remain so for decades, at least. But one can be confident that human ingenuity will one day enable such a concept to enter the realms of possibility, for instance if the function of a transistor can finally be implemented by a single molecule, organic or not. In such a case it would be conceivable to construct a network of artificial neurons containing the hundreds of millions of components of the human brain.

The Essential Question

Depending on whether such a brain would reproduce, or not, all aspects of the behavior of its human prototype, the fundamental philosophical question that opposes materialists and dualists would then be answerable experimentally. We have evoked this question since the beginning. Now is the time to ask it again, and be specific.

According to the materialist hypothesis, as defended by English pioneer of DNA research and Nobel Prize winner Francis Crick (1916–2004), the material world is closed. It depends on brains that perceive signals from the senses, that memorize them in part, that organize themselves in terms of the same outside stimuli, and then control the actions of the body. No subjective, mental experience exists outside this material substrate, that is modifiable by this same experience. Thus man's supposed enjoyment of free will is entirely determined by the physical activity of his brain, and is illusory. According to this hypothesis an artificial brain should function like a natural brain. If such a brain could be built it should be possible to demonstrate that man is in fact a robot. And if we could build a robot that behaves like a man, we could prove that that man is merely a robot.

In the dualist hypothesis the brain is controlled by an immaterial spirit that is the seat of subjective, mental experience. The brain is merely an interface used by this spirit to acquire information about the outside world and transmit orders to our motor organs. According to this hypothesis an artificial brain would not function at all like a natural brain, or it would demonstrate its limited character. This dualist thesis is a frontal challenge to the principle of the conservation of energy in supposing that certain neurons are activated by orders from a spirit even when there are no incoming signals from other neurons. But it is the only thesis that preserves any autonomy or liberty for the human mind such as we imagine it subjectively. I exist in as far as I make decisions; whether these decisions are the result of mature reflection or founded on some deep intuition, I am free.

According to classic physics, the whole universe is an automaton of which the future activity is entirely determined by its present state. This rather summary idea can be traced back to French mathematician Pierre-Simon Laplace (1749–1827) and it inspired much of the scientific reflection of the nineteenth century. It is therefore not surprising that it spawned the materialist hypothesis concerning the relationship between brain and mind.

In fact, we have since discovered that the physical world is not as deterministic as Laplace's theory might lead us to believe. On the one hand, since the beginning of the twentieth century, quantum mechanics has introduced the concept of an essential uncertainty in the behavior of a physical system, as long as the phenomena involved are considered at the microscopic level. In addition, phenomena of chaos, mentioned earlier, are of a deterministic nature although they remain unpredictable: their sensitivity to the initial conditions is such that it is not possible to calculate their outcome.

Australian neurophysiologist, and Nobel Prize winner, John Eccles (1903–1997) defended a dualist hypothesis concerning the brain. The influence of an immaterial spirit on a synapse would signify a material movement at a sufficiently microscopic level that it would effectively involve an order of magnitude where the uncertainty principle of Heisenberg would apply. Random fluctuations would be such that conservation of energy would only be respected on average. Nothing prevents three quanta of energy from activating the first neuron in a chain without violating any law of physics. Scientifically we cannot demonstrate, or refute, the hypothesis of an immaterial spirit acting on the brain.

Furthermore, there is now systematic research on chaos related to brain activity. The appearance of chaotic phenomena in artificial neuronal networks is obvious, for they are non-linear circuits of adequate size and subject to adequate excitation. They are even phenomena that can be observed in any binary robot, such as a modest logical circuit with three or four gates for example.

The pursuit of research in these two directions will doubtless allow us to better address the universal problems created by the historical emergence of the human mind. It would be presumptuous to anticipate the result. It would be just as illusory to imagine today that the philosophical problem evoked above will one day have a scientific solution, in the strict sense of the term. All that physics or technology can offer in this domain is of the order of negatively metaphysical: they can eliminate certain solutions, but cannot prove any. They can dispel a few simplistic ideas about the existence of an immaterial spirit, but cannot penetrate its true nature for, by definition, it is outside the field of experimental observation.

Eccles' thesis is doubtless the result of a dualist conviction that drove him to search for a physical explanation. One can understand its origin and its outcome. But it does not deserve to be neglected. The truth is much more complicated that we might imagine. Although quantum physics is about a century old now, it still has not been integrated into our ordinary vision of nature. It describes an extraordinary world of effects without causes, where chance is master, and where uncertainty rules over even the position of particles. What is our vision of the universe worth? Is it not a creation of our own brain? Dualism is certainly an archaic stance to hold, but materialism is doubtless no less so.

What poorly understood or totally unknown physical effects might explain certain marginal phenomena, ignored by the mainstream of scientific research, such as telepathy and clairvoyance? Such questions animate the Odier Foundation for psychophysics, based in Geneva. These psychophysical phenomena are little studied by most laboratories that are either frightened of harming their serious reputation, or of getting lost in research on non-reproducible random phenomena, that are by their very nature excluded from classical science. Among others, the foundation has subsidized research in the laboratory of Nicolas Gisin at the University of Geneva. He has devoted much time to the experimental verification of the Einstein, Podolski and Rosenberg (EPR) theory and quantum teleportation. Without going into more detail, his results show that information can be very rapidly transmitted over long distances without an electromagnetic support, that we know to be relatively slow. We thus cannot exclude that our brain perceives and processes information according to physical laws that are unfamiliar or unknown at the moment.

Answers to the Essential Question

Can a machine simulate the highest cerebral functions to the extent of becoming indistinguishable from a human brain?

If you ask this question to a selection of neuroscientists you obtain very varied replies. It is soon obvious that it is of little concern to them, for they have never really asked it of themselves. They do not think about it, it disturbs them, and their replies are dictated by pure intuition.

Many people see the future in the light of a catastrophe. Ray Kurzweil, American inventor, and author of *The age of spiritual machines* (1999), has predicted that at the end of the twenty-first century computers with capacities 100 thousand times more than man's brain will run the world. For the first time on Earth there will be predators with more developed mental capacity than man. Just as man once dominated the world, they will dominate, and man will be in zoos. Whether there is a difference between a biological brain made of carbon and an electronic brain made of silicon is a non-question and a non-problem, for the two already interact and will end by fusing.

Engineers and Biologists Afflicted with the Karl Marx Syndrome

When philosophers became aware of the existence of the steam engine and its decisive influence on society they settled into a period of reflection, in their own peculiar way, with disastrous results. History, or maybe legend, recounts that Karl Marx never worked in a factory and that he never even visited one. His office-bound reflections on the industrial revolution led to devastating effects when communist regimes set their minds to applying the remedies proposed by an intelligent man who did not know what he was talking about.

It is therefore not without interest to wonder if engineers and biologists are adequately equipped to investigate the problem of consciousness. If we are trying to elucidate the various mechanisms that make us visually conscious of our environment, that is not too difficult. Scientists will end up by understanding how the different cortical areas communicate to provide us with an integrated consciousness of objects, with their outlines, colors, spatial positions, speed and so on. It is reasonably clear that visual reality does not exist in our consciousness other than as a result of neuronal processing of signals perceived by the retina. In view of the more or less serious imprecision in this processing, we can easily imagine that we might see nonexistent objects and not see existing ones.

But our achievements may conceal a trap: the bioengineers of consciousness run the risk of catching the Karl Marx syndrome. Reality would be reduced to what their methods allow them to comprehend. They might just about admit that certain phenomena could escape their analysis, but that these would be of little impact, being impossible to model.

It is a pity that Marx did not make the effort to better adapt his technique by working in a factory. His contribution would have been richer, more pertinent and less dangerous. We might similarly prefer that the engineers and biologists who study brain function make a serious effort to be familiar personally with certain experiences that their supposed workload might induce them to underestimate: listening to music, contemplating pictures, falling in love, praying, and, above all, feeling pleasure through these activities. They might then attach more importance to such affective events and less to the cortical area that each excites. They would have the time to concentrate on their impressions and enjoy them at leisure.

The story of the murder of Guillaume Apollinaire's passion

In 1914 Guillaume Apollinaire, whose real name was Apollinaris de Kostrowitsky, was 34, but was already considered to be one of the most important French poets. He volunteered to serve in the First World War, was promoted to second lieutenant, and was wounded in the head on March 17, 1916. Three days after his wound he was transferred to the Val-de-Grâce Hospital in Paris where he was treated by a highly competent neurologist. She noted a superficial lesion of his right temporal lobe, but an absence of symptoms: she detected nothing abnormal in terms of vision or speech, no loss of consciousness, no motor troubles. She did however notice Apollinaire's disheveled appearance.

The only notable consequence of his brain injury was a sudden intolerance of any form of emotion. Whereas before he had held a passionate correspondence with his fiancée, Madeleine, his letters after his injury are banal and cold. He asked her not to write to him again and not to visit him: he had become incapable of accepting emotion. It was the same with his "war godmother" Jeanne Burgues (the poetess "Yves Blanc") to whom he confided that he had become very nervous and excessively irritable.

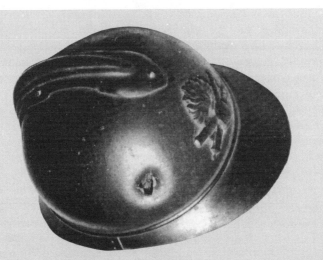

Figure 11 Guillaume Apollinaire's helmet with a two-centimeter-square hole caused by a fragment of shrapnel

Guillaume Apollinaire broke off his relationship with Madeleine without any explanation, and he never saw her again. He became anxious and tormented, as we can see in the portraits of him by Jean Cocteau and Jean Hugo. In contrast, he suffered no loss of memory or creativity, and continued to write, even if his later poems reveal deep sadness. So there was complete dissociation between his social and affective behavior on the one hand and his poetic gifts on the other. This syndrome would be recognized today as that of a lesion in the right temporal lobe, with preservation of all other regions. A similar case was to occur a decade later with George Gershwin who suffered a tumor in the same region: he underwent psychotherapy for his changed personality. Apollinaire's case was also diagnosed at the time as an episode of depression as a result of the psychological trauma of his wound. In fact, he stopped being in love just because of the loss of a few neurons in his right temporal lobe.

Chapter 11
The Power and the Fragility of Oneself

The pineal gland, so dear to René Descartes in the seventeenth century, was for long relegated to the rank of an accessory organ by all who studied the brain. This accessory organ is in fact essential, for it regulates sleep. But it is useless to search in it, or elsewhere, the link between an immaterial spirit and the brain. The lofty dualism of the philosopher, the common dualism of everyday language, the reasoned dualism of the casuists, the lazy dualism of backward scientists, the fascinated dualism of amateurs of all that is marvelous, all these tools of western thinking, are to be put away in the dusty drawers of outdated theories. They are based on summary ideas that are the first to occur to one's mind because they seem obvious. We never managed, and we still do not manage, to accept that thought, in all its splendor, is produced by a liter and a half of protein organized by a process of biological evolution. We prefer, as in children's fairy tales, to think of ourselves as ideas cast in matter, angels imprisoned in a body that is more encumbering than useful.

The End of Vitalism

Moreover dualism is marked by a contradiction, like that of vitalism. For a long time it was believed that living matter was essentially different from inanimate matter in that is was inhabited by a vital principle that was so subtle that it was impossible to synthesize in the laboratory. But with the discoveries about DNA, especially those of Francis Crick and James Watson in the 1950s, and the more recent elucidation of the human genome, the idea was eclipsed. We do not know all the processes that regulate the function of the cell. We would be incapable for the moment of reconstructing one from its chemical constituents, and we can only vaguely imagine the phenomena that prompted its first appearance. But still we have abandoned the vitalist theory in the scientific world.

Life does not depend on a subtle principle which is immaterial because we cannot identify it, but nevertheless has a physical reality. It is not, though, like that for much of the general public. Their reaction of panic when confronted with progress in biology shows that. People do not much like human DNA profiling, genetically modified plants, the use of embryonic stem cells, or preimplantation

J. Neirynck, *Your Brain and Your Self: What You Need to Know*,
© Springer-Verlag Berlin Heidelberg 2009

diagnosis as part of medically assisted procreation. The phenomenon of life is surrounded by a sacred mystery based on fundamental ignorance. People understand easily enough that a dead cell decomposes into inanimate material, but they find it difficult to accept the opposite, that is to say that life stems from the organization of molecules and not from any change in their nature, if the word nature has any sense here.

The organization of a molecule of DNA alone explains the mysteries of living matter: its ability to replicate, its way of absorbing its very substance from its environment, its diversity within the same organism, its evolution, and finally its mortality. Inanimate matter possesses the property of becoming animate. Life is an emergent property of matter: providing that the necessary materials are available, and that environmental conditions are favorable, living matter will emerge.

The marvel of life is not diminished by this. At the moment of the Big Bang, the primal soup of quarks, devoid of any diversity or interest, possessed the property of condensing into electrons, protons and neutrons, then into the atoms of Mendeleev's periodic table, then into molecules and finally into DNA. Such is the plan of the creation of the universe as we can conceive it today: it is no less marvelous than those imagined by our ancestors in the fundamental myths of our culture.

The Trap of Reductionism

Confronted with the brain, we are placed before the same cultural leap. We must now admit that thought emerges from the structuring of cellular life, just as the living cell emerged from the structuring of organic molecules. We have had to face this from all the cases spelled out in this book: the mind depends on the brain, and small changes in the brain spell considerable consequences for human mind.

We can admit it fairly easily in the case of Parkinson disease: a lack of dopamine, an essential neurotransmitter, the nerve impulse is badly transmitted and the patient's whole motor skill is affected. An intact mind is held prisoner in a body that is the victim of slow paralysis. But we are uneasy when Alzheimer disease destroys a person's neurons to such an extent as to annihilate his previous personality. He declines into a body in which the mind diminishes until it disappears. So our natural inclination toward dualism makes us prefer a mind prisoner of a defective body rather than a lost mind in a healthy body. If the mind no longer resides in a body, should this body not simply die? If man is equal to a body plus a soul, the body being destined for the grave and the soul for eternal life, what is the point of a living body with no soul? Where has this soul gone?

With this concept of the fragility of our being, the danger is to sin by excess in the opposite direction, the direction of a doctrinaire materialism, a reduction of our higher faculties to simple combinations of neurons. But our affectivity, our esthetic sense, our artistic or scientific creativity are all realities that are as respectable as the functioning of a neuron in an MRI machine. Man's mind is just as objective a reality as the existence of cells.

Reductionism is an extreme oversimplification. It is scrutinizing the trees and forgetting the forest. It is not because the mind does not exist outside the brain that it does not exist at all. But not in the sense of some immaterial vapor floating above the body during an out-of-body experience, or becoming independent and almost attaining immortality during a near-death experience. To look for the seat of consciousness, the physical link between mind and body, in the framework of dualism implies that in fact you remain at the level of materialism: the mind would exist because you would have surveyed it, measured it, weighed it, trapped it by some conclusive and definitive experiment.

It is possible to develop a conception that avoids the symmetrical traps of dualism and materialism, providing that one agrees to reflect on the significance of models of natural phenomena and the need to use several simultaneously. For, to imagine that man is both body and soul, material and immaterial, mortal and immortal, is to once again formulate a verbal model to explain our experiences. The first thing is to overcome the simplistic idea that a single model represents reality. To exploit scientific method, which always implies a phase of modeling, one must conceive as many different models of reality as necessary both for reflection and action. A model is useful in so far as it is intelligently remote from reality, because its aim is to simplify reality to such an extent that we can appreciate it. We are not involved in a competition between models, in which the best would be the closest to reality, and thus the most complicated. The methodology of scientific modeling is far from obvious and it gives rise to much misunderstanding, even among scientists.

A Model is not Reality

Let us explain this with the example of a model which we use in everyday life, a topographical or road map. The most detailed map of a given piece of land is the surveyor's map, for every property is drawn with its designated limits and the buildings that stand on it. It is probably the closest to reality, but the leisure walker would not encumber himself with this sort of map, for it would entail a sheaf of papers if his aim was to walk along roads and lanes, and even across land without designated pathways.

Topographical maps at a scale of 1:10,000, which means a centimeter on the paper represents 100 meters in reality, would be more useful to him, because they portray the details needed to find his way through the countryside, such as contours, the nature of the terrain, fields, woods or built-up areas, but especially because they leave out details which, although interesting, are irrelevant to the walker. However, even the most detailed map omits what is the essential for a successful stroll: the beauty of the landscape, the song of the birds and the perfume of the flowers.

But this indispensable tool for the pedestrian becomes in turn of little use to the motorist, restricted to negotiable roads, for whom such detail is irrelevant and who prefers maps on which a centimeter represents several kilometers. On the other hand, the type of road is of importance to him: he wishes to be able to distinguish between a highway and a gravel track. Indeed, the driver who simply wishes to cross the countryside as quickly as possible will find a map showing only highways quite sufficient.

Furthermore, it is not just the scale of a map that is important. We produce maps showing geology, politics, demography, linguistics and sociology, each of which concentrates on a certain type of information. The common feature of all these maps is not how close they are to the reality of the landscape, but a degree of remoteness from reality calculated to emphasize only the prime interest of a given user, and to eliminate the superfluous.

Whatever features are represented on a map, it cannot replace reality. One could consider France as a territory of 551,602 square kilometers, at a mean altitude of 342 meters, situated between 42° 20' and 51° 05' latitude north, inhabited by a host of plants and animals, including man. If we want to be even more precise we could make a chemical analysis of the soil and record the atoms of silicon, carbon, nitrogen and hydrogen. The more we enter into detail, the less we perceive the most important vision: the French nation with its culture, its history and its institutions.

The natural sciences work like cartography. They imagine models that do not cover the whole of reality, but which represent a certain aspect, at a certain level, without claiming to enter into all details. This what the Blue Brain project, mentioned in the previous chapter, does by producing a mathematical model of a cortical column composed of thousands of neurons from measurements on the rat brain. This model is both very detailed and rather big, more so than anything achieved until now. But it does not represent reality at the molecular level or even that of DNA, and even more so at the much more "microscopic" level of the atom or the elementary particle. If it did so it would lose its usefulness, because it would become so complicated that the calculations would be of inordinate proportions. The talent of a researcher is to conceive a model that is sufficiently close to reality to explain the phenomenon being studied, but astutely remote to exclude irrelevant phenomena. So we can conclude that even by adding together a million cortical columns modeled at the neuronal level we could not reproduce reality, because that is impossible. The only model that is totally faithful to reality is reality itself. All other models simplify, schematize, neglect. To encompass the mind in a neuronal model may not be impossible, but cannot be guaranteed, and it does not seem to be the best model to do this. Markram's research seeks to unravel the organization of a tiny part of the brain, and not to simulate it entirely. One can therefore understand the reticence of certain scientists when asked if a neuronal simulator will one day be indistinguishable from a human brain.

Existence of the Mind

The mind is a convenient concept for modeling the psychic phenomena that make the richness of man. It is studied in psychology using the methods of the social sciences, without relying on cerebral imaging or analysis of cerebrospinal fluid. Psychotherapy, attempting to relieve psychological deficits by using words without necessarily having recourse to medication, is a legitimate principle, even if sometimes its practice can be abusive.

If a patient suffers from an obsessive-compulsive disorder as a result of his upbringing, life events or verbal experience, it is possible to unravel the maze by word of mouth, without however excluding a treatment based on antidepressors. The experiences we live program us physiologically by reinforcing or weakening our synapses. Psychotherapy can work in the opposite direction without necessarily having recourse to chemical treatment. What words have made wrong, words can put right.

To educate a child, we use oral or visual processes and repetition for learning to write, to read and to calculate: we make connections in the developing brain by considering it at a global level without concerning ourselves with the details of the synapses. We do not hand out pills in class to improve memory. We only interfere at a chemical level in case of pathology.

So, the mind forms a real entity, sufficient for processing at a certain level of reality. It exists, neither more nor less than a glial cell or a molecule of dopamine, both of which are needed for its proper functioning. It is a map of the person at the most abstract level: it is most appropriate in circumstances such as social relations. If a patient develops an obsession, such as anorexia, it is impossible to reprogram the millions of synapses where it has gradually been expressed. The disease affects a global cerebral function that involves widespread regions of the brain, accessible globally, and reprogrammable by sessions of psychotherapy, perhaps accompanied by medication.

Furthermore, the brain does not exist independently of the body. It changes its structure by exploring the environment of the body. What will always be missing from an ordinary computer is the experience of a body and it is just this obstacle that Braitenberg's research tries to avoid.

But there remains a question: is the concept of the mind as a model of brain and body at the highest level, not merely an abstraction, a metaphor, a fiction? Is this an indirect way to reintroduce a form of dualism purged of its archaic origins? We would like a confirmation of the power of the mind, its power to command the neuronal machine. One can find it in the phenomenon of the placebo.

Proof by Placebo

The word placebo is from the Latin for "I please", meaning "I please the patient". It designates those remedies devoid of specific biological activity, such as sugar or salt water. We know by experience that a placebo often has an effect. "Often" means, for instance, in about 10% of cases for Parkinson disease and 90% for arthritis. "Often" also varies with the method of administration: an injection is more effective than a capsule, which is more effective than a pill, which is better than a liquid medicine. Furthermore, a placebo is more effective when given in an aggressive hospital environment, rather than in a more homely atmosphere. "Often" also varies with the color of the medication. And effectiveness increases when the prescriber is "blind" to when a placebo is given. The questions that arise are to know if the patient is really cured, if he is simply pretending he is cured, or if he was in any case suffering from an imaginary illness. In the last case we have to distinguish between the patient simulating a condition without being aware of it, and he who is simulating in order to manipulate his entourage.

One of the key passages of Molière's comedies in the seventeenth century is the hilarious dialogue between the imaginary invalid and the charlatan doctor: the two protagonists pretend to believe in the efficacy of absurd remedies in order to deceive each other and to manipulate their entourage. In the end, everyone is deceiving himself and finishes by believing in the effectiveness of the prescriptions. Up to the nineteenth century much of medicine relied on placebo-style remedies. This form of treatment would not have survived had it not led to real cures. Indeed we may ask the question today as to whether this is not the explanation and justification of homeopathy, acupuncture and the laying-on of hands.

Not all illnesses respond to a placebo. It is not effective in organic diseases like infections and cancer. In contrast a placebo can triumph over functional diseases such as depression, gastric ulcer and hypertension, and symptoms such as cough or pain. So much so that one must take this into account when validating pharmaceuticals intended for use in these illnesses. To avoid the "placebo effect", one uses a "double blind" trial: patients are selected randomly into two groups who receive either the drug under study or a placebo, without their knowing which group they are in. In addition, to make this process completely rigorous, the physicians who make the diagnosis, prescribe the drugs and evaluate the results do not know which patients receive the medication under trial and which the placebo. In other words, the placebo effect is so real that a complex strategy is necessary to allow for it and reveal any effective action of the medication.

It is easy to see that such a procedure raises a fundamental problem: what is this placebo effect that appears practically always under certain circumstances? An illusion by the patient? It was not fully realized that a placebo has a real physiological basis until 1978. After the extraction of a tooth, Californian neurologist John Levine found he could effectively relieve the very real pain either by giving morphine, or a placebo. He suspected that the placebo suppressed the pain by producing endorphins in the body of the patient. To verify this, he administered to the placebo patients a drug that blocked the action of endorphins: the pain came back. So it is not just a purely psychological phenomenon of suggestion, without physical support. The mind, influenced by a psychological manipulation, actually orders the body to take the necessary physiological action.

In 2002, Martin Ingvar and his colleagues in Stockholm, using PET imaging, discovered that a placebo activates the same regions of the brain as opioid pain killers, of the same essential nature as endorphins. In 2001 Jon Stoessl in Vancouver conceived the hypothesis that a placebo acted on Parkinson disease by producing dopamine in the brain. He also used PET and found that, indeed, injection of saline produced dopamine in the body. In 2002 Helen Mayberg in Atlanta showed that Prozac, the well-known antidepressor, can be replaced by a placebo and that both treatments are associated with increased activity in the cortex and decreased activity in parts of the limbic system. The only difference was that Prozac's effect lasted longer.

All these convergent experiments demonstrate that there is a physical basis for the placebo effect: it is not merely a figment of the imagination or a desire of patient and physician to please each other. The mind, feeling that it has received an effective treatment, stimulates the same body mechanisms that would have been stimulated chemically by a specific drug. A physical result stems from a psychic impulse. How? For the moment we do not know.

In practical terms, we must act not only as if the mind existed, but must admit that it really exists. Just as it depends for its existence on a healthy brain, this abstract element can act in the opposite sense, that is on the brain itself. It is not necessary to imagine it like a cloud floating in some other dimension that our senses cannot access. The mind exists because of how the brain and the rest of body are organized: it is an emergent property of the organization of living matter.

So we have disposed of Bergson's thesis. We might wonder how, at the turn of the nineteenth and twentieth centuries such a cultured man, familiar with the top discoveries of contemporary medicine, living in the center of the scientific world that was Paris at the time, could have ignored the discovery of his colleague Broca, who showed without a shadow of doubt that a circumscribed brain lesion was associated with loss of a specific mental faculty. His spiritualistic penchant obscured his discernment. But it is not necessary to take refuge in a dualist utopia to believe in the existence of the mind. Does it survive death? The reply to this question depends on intuition, belief, and religious faith. It does not depend in any way on a philosophical theory that has been destroyed by scientific experimentation.

This leads to a question that has serious consequences: do we dispose of free will? Are we able to make personal decisions, in trivial or serious circumstances, that are not just the automatic result of the configuration of our synapses and the sensory information that we receive? Or are we robots that believe, wrongly, that we make decisions? Are we free to make vital decisions, crucial for our life and even our very existence, such as choosing to study, or selecting a profession, whether to marry, commit a crime or not, fight an illness or commit suicide?

Opinions on Free Will

In murder trials there is often evidence from one or more psychiatric experts who give their opinion on the criminal responsibility of accused. Was his act carried out after mature deliberation by a mind capable of free choice? Was he capable of not committing his crime if he had inhibited his murderous impulse by a conscious decision? We know how such diagnoses are uncertain and contradictory. They stem from the dualist illusion according to which a human being in good mental health can decide freely.

Even without considering such a dramatic situation as murder, we may still ask the same question about everyday life. Each day we make a host of decisions, and we believe fervently that they are the result of our own free will. We decide when we are going to shake off the night's slumber, get up and face the day; whether we are going to have a drink at lunchtime, when we know we would be better off not to; whether to daydream or to get on with some urgent work; whether to waste an evening in front of the television with a beer, or indulge in some culture by reading a difficult book. We often feel that we ought to be making some important decision with no further ado, because by dragging our feet we may well make a wrong decision out of pure laziness. Sometimes we make a courageous decision out of the blue. Just what is going on? Is there an immaterial "self" making these decisions by activating the first neuron in a chain that excites others one by one until the motor cortex is triggered to work our muscles? If such is the case, what has happened to the sacrosanct principle of conservation of energy? Where does the first spark come from that sets off the first neuron?

If I do not exist "myself" it implies that all my actions are decided by neuronal connections and sensory inputs. I act like a robot, but one that is too complicated for you to be able to predict my behavior, and I can therefore hide behind your ignorance so that I seem to be making decisions.

Several experiments support such a thesis. In 1983 Benjamin Libet (1916-2007) and his colleagues in San Francisco described a protocol in which he asked a subject to indicate very precisely the moment he decided to move a finger. At the same time Libet measured the electrical activity of the motor cortex. He discovered that his subject was conscious of the intention to make a movement 350 milliseconds **after** the start of the electrical activity, the so-called readiness potential. From this experiment, Libet suggested that our free will was limited to being able to inhibit already planned activity.

Still more striking, recent experiments show that our perception of the time separating two events depends on the idea that we have of the causal link between the two events. The time seems shorter than it really is if we perceive the second event as a consequence of the first, deliberate event. This illusion of the relationship between our intentions and our actions might be the mechanism that the brain uses to give us the impression that we control certain of our acts.

In summary, the brain takes note of a decision to act while the neurons preparing the act have already been firing for some time. We are not conscious of this situation. We think we make a decision, but in fact we simply become aware of an act undertaken by a robot.

We are not trying to settle this question here, but it is of interest to expose the dilemma of free will to scientists working in the field. In their model of the brain, in the perception they have of it through their research, does free will exist or is man, without being aware of it, merely a robot? Their answers vary, but most seem to grudgingly admit that the subjective impression of a degree of free choice in life is necessary at the subjective level.

But we leave the last words, as we did the opening words of this book, to Patrick Aebischer: "I try to identify those aspects of life about which we can speak from experience. About those we cannot experience, it is better to say nothing, as Wittgenstein said so well. So as a scientist, I have nothing to say about free will"

Bibliography

The following references are given as a few suggestions of books, articles and Internet sites that may illustrate and further develop some of the points made in the text. They are by no means comprehensive.

Books and articles

Bear MF, Connors BW, Paradiso MA (2006) Neuroscience: Exploring the Brain. Third edition. Lippincott Williams & Wilkins, Baltimore

Bogousslavsky J, Boller F (eds) (2005) Neurological Disorders in Famous Artists. Frontiers of Neurology and Neuroscience. Vol 19. Karger, Basel

Braitenberg V (1984) Vehicles: Experiments in Synthetic Psychology. MIT Press, Cambridge

Brodmann K (2006) Localisation in the Cerebral Cortex. Third edition. Translated by L Garey. Springer, New York

Brody H (2000) The Placebo Response. Harper Collins, New York

Chakradhar ST, Agrawal VD, Bushnell M L (1991) Neural Models and Algorithms for Digital Testing. Springer, New York

Changeux JP (1997) Neuronal Man: The Biology of Mind. Third edition. Translated by L Garey. Princeton University Press, Princeton

Changeux JP, Ricoeur, P (2000) What Makes Us Think? Translated by MB DeBevoise. Princeton University Press, Princeton

Crick F (1994) The Astonishing Hypothesis: The Scientific Search for the Soul. Macmillan, New York

Damasio AR (1994) Descartes' Error: Emotion, Reason and the Human Brain. Harper Collins, New York

Dawkins MS (1998) Through Our Eyes Only? The Search for Animal Consciousness. Oxford University Press, Oxford

Dayhoff JE (1990) Neural Network Architectures: an Introduction. Van Nostrand Reinhold, New York

Dennett DC (1991) Consciousness Explained. Little, Brown, Boston

Duke D, Pritchard W (eds) (1991) Measuring Chaos in the Human Brain. World Scientific, Singapore

Eccles JC (1991) Evolution of the Brain: Creation of the Self. Routledge, London

Evans D (2003). Placebo: Mind over Matter in Modern Medicine. Harper Collins, New York

Gelenbe E (ed) (1991) Neural Networks: Advances and Applications. North Holland, Amsterdam

Greenfield S (1995) Journey to the Center of the Mind. WH Freeman, New York

Hertz J, Krogh A, Palmer RG (1991) Introduction to the Theory of Neural Computation. Addison-Wesley, Redwood City

Hopfield JJ (1982) Neural networks and physical systems with emergent collective computational abilities. Proceedings of the National Academy of Sciences 79 2554–2558

Jouvet M (1999) The Paradox of Sleep: The Story of Dreaming. Translated by L Garey. MIT Press, Cambridge

Kandel ER, Schwartz JH, Jessell TM (2000) Principles of Neural Science. Fourth edition. McGraw-Hill, Columbus

Kurzweil R (1999) The Age of Spiritual Machines. Viking, New York

Lisboa PJG (ed) (1992) Neural Networks: Current Applications. Chapman and Hall, London

Moody R (1975) Life after Life. Mockingbird, Covington

Nicholls JG, Martin AR, Wallace BG (2001) From Neuron to Brain. Fourth edition. Sinauer, Sunderland

Popper KR, Eccles JC (1977) The Self and its Brain. Springer, Berlin

Sacks O (1985) The Man who Mistook his Wife for a Hat and Other Clinical Tales. Summit, New York

Sanchez E (ed), (1992) Artificial Neural Networks. IEEE Press, New York

Shepherd G (1994) Neurobiology. Third edition. Oxford University Press, New York

Web sites

General information:
http://www.brainexplorer.org/
http://thebrain.mcgill.ca/
http://www.sciam.com/sciammind/
http://web.mit.edu/bnl/

Korbinian Brodmann:
http://l.garey.googlepages.com/brodmann.pdf

Phineas Gage:
http://www.deakin.edu.au/hmnbs/psychology/gagepage/

HM and memory:
http://homepage.mac.com/sanagnos/scovillemilner1957.pdf
http://web.mit.edu/bnl/pdf/Corkin2002.pdf

Brain diseases:
http://www.healthfirsteurope.org/index.php?pid=75

Parkinson disease:
http://www.pjonline.com/Editorial/20000226/education/parkinsons1.html
http://www.pjonline.com/Editorial/20000325/education/parkinsons2.html
http://www.ninds.nih.gov/disorders/parkinsons_disease/detail_parkinsons_disease.htm)

Alzheimer disease:
http://www.pjonline.com/Editorial/20000603/education/alzheimers.html
http://en.wikipedia.org/wiki/Alzheimer%27s_disease#Biochemical_characteristics
http://en.wikipedia.org/wiki/Memantine

Brain tumors:
http://www.cancer.gov/cancertopics/pdq/treatment/adultbrain/HealthProfessional
/285.cdr#Section_285

OBE and NDE:
http://brain.oxfordjournals.org/cgi/reprint/127/2/243
http://profezie3m.altervista.org/archivio/TheLancet_NDE.htm

Ray Kurzweil:
http://www.kurzweilai.net/meme/frame.html?main=/articles/art0274.html

Index

Printing: Krips bv, Meppel, The Netherlands
Binding: Stürtz, Würzburg, Germany